MEMORIES OF THE OLD HOME PLACE

MEMORIES OF THE OLD HOME PLACE
Prince Edward Island

✤

Text by
Francis W. P. Bolger

Photography by
Wayne Barrett & Anne MacKay

ACKNOWLDGEMENTS

We thank Kimberley Reid for her assistance with the research, Shirley Dillon for her preparation of the typescript, and Frank Ledwell for his reading of the manuscript. We thank the following Island residents for their help and information: Allan and Pearl Barrett, Austin Bolger, Ernest Bolger, Rev. Austin Bradley, Buddy Campbell, George Campbell, Norman Campbell, Ruth Campbell, George Carr, Stan Carr, Rev. Robert Coady, Laurie and Sally Coles, Lester Gavin, Goldie M. Gillis, Donald Harris, Harry Harris, Donald Harron, Rev. Preston Hammill, John and Jean Haslam, Richard S. Hinton, Leonard Hogan, Gary and Roger Hutt, Earl Larkin, Sarah Lea, Malcolm Lodge, Florence McCarey, Barbara MacDonald, Jospeh G. MacDonald, Judy MacDonald, Leslie MacDonald, William J. MacDonald, Jean MacFadyen, Rev. Wendell MacIntyre, Donald and Audrey MacKay, Eric Ike Moreside, Kevin O'Neill, Sandy MacNeill and family, Regan Paquet, Franklin Perry, Allan Rankin, Mary Cullen Raymond, Irene Rogers, Raymond Sark, Rupert and Vera Simpson, Stanley Skeffington, Rev. Robert C. Tuck, and David Weale. We also thank the following bodies for their cooperation and permissions relative to the photographs: The Prince Edward Island Museum and Heritage Foundation, The Confederation Centre of the Arts, CN Marine, The Prince Edward Island Department of Tourism, *Les Danseurs Evangéline*; The Confederation Centre Summer Festival for permission to reproduce a photograph of the musical *Anne of Green Gables*, and Morning Music Ltd. for permission to reprint the lyrics of Stompin' Tom Connors' 'Song of the Irish Moss'. Finally we especially thank Ruth L. Paynter for permission to portray her quilt on our book jacket and to use its title as the sub-title of our book.

F.W.P.B. W.B. A. MacK.

OXFORD
UNIVERSITY PRESS

70 Wynford Drive, Don Mills, Ontario M3C 1J9
www.oupcan.com

Oxford University Press is a department of the University of Oxford. It furthers the University's objective of excellence in research, scholarship, and education by publishing worldwide in

Oxford New York

Athens Auckland Bangkok Bogotá Buenos Aires
Calcutta Cape Town Chennai Dar es Salaam Delhi Florence
Hong Kong Istanbul Karachi Kuala Lumpur Madrid
Melbourne Mexico City Mumbai Nairobi Paris São Paulo
Singapore Taipei Tokyo Toronto Warsaw

with associated companies in Berlin Ibadan

Oxford is a trade mark of Oxford University Press in the UK and in certain other countries

Published in Canada
by Oxford University Press

Copyright © Oxford University Press Canada 2000

The moral rights of the author have been asserted

Database right Oxford University Press (maker)

First published 1984 (hardback)
First published 2000 (paperback)

All rights reserved. No part of this publication may be reproduced, stored in a retrieval system, or transmitted, in any form or by any means, without the prior permission in writing of Oxford University Press, or as expressly permitted by law, or under terms agreed with the appropriate reprographics rights organization. Enquiries concerning reproduction outside the scope of the above should be sent to the Rights Department, Oxford University Press, at the address above.

You must not circulate this book in any other binding or cover and you must impose this same condition on any acquirer.

Canadian Cataloguing in Publication Data

Bolger, Francis W.P., 1925-
Memories of the old home place : Prince Edward Island

Includes bibliographical references.
ISBN 0-19-541511-6 (pbk.)

1. Prince Edward Island – Description and travel. I. Barrett, Wayne.
II. MacKay, Anne. III. Title.

FC2617.4.B64 2000   917.17'044   C84-098656-4
F1047.B64 2000

Produced by Roger Boulton Publishing Services, Toronto
Designed by Fortunato Aglialoro

1 2 3 4 - 03 02 01 00

This book is printed on permanent (acid-free) paper ∞
Printed in Hong Kong

INTRODUCTION

A few years ago, *The Charlottetown Guardian* published a letter that a young Charlottetown student had received from the Empress Hotel in Victoria, BC, in response to her application for a position. The letter, from an executive officer of the Canadian Pacific hotel, thanked the student for her interest but pointedly added: 'Unfortunately we are not in a position to offer employment to those residing outside of Canada.' While the newspaper dutifully expressed concern that Prince Edward Island obviously formed no part of CP's Canada, Islanders throughout most of their history would have revelled in this recognition of their highly prized 'splendid isolation.'

Prince Edward Island, a picturesque, crescent-shaped island, 140 miles in length and varying from two to thirty-five miles in width, is cradled peacefully in the Gulf of St. Lawrence, a few miles off the shores of New Brunswick and Nova Scotia. Over the past hundred years the population of the Island increased only to its present 126,000 from 108,000 a century ago, and it had actually declined, meanwhile, to a low of 88,000 in 1931. Some years ago these startling statistics led the president of an over-populated African republic to write to the premier of the Island asking for a copy of the province's legislation on family planning! Those closer to the scene recognize that, in spite of a high birth rate, the Island population has remained relatively stable because of the heavy emigration necessitated by limited land for agriculture and the lack of significant secondary industry. But wherever Prince Edward Islanders may settle, they remain Islanders. They speak of their home as 'The Island' without qualification, explanation, or apology. To Prince Edward Islanders, the Island is not so much a geographical entity as a way of life.

A strong sense of identity has always been the distinguishing characteristic of the history of PEI. Both the Micmac Indians and the Acadians during their periods of dominance considered the Island *their* home and themselves an independent people. When Great Britain assumed control of the Island in 1763 and annexed the island—called the Island of St John— to Nova Scotia, she nevertheless soon realized that the Indians and Acadians, although dispossessed and impoverished, had a sense of identity that should be respected. Moreover, even the approximately one hundred proprietors to whom the British government had callously granted sixty-six of the sixty-seven townships on the Island demanded immediately that it should have independent status. In 1768 a large number of these proprietors petitioned the king of England, stressing that Halifax was too far removed from Charlottetown to serve as a satisfactory capital and that the settlement of the Island would proceed more expeditiously if it were separated from Nova Scotia. The king accepted their advice, and on 28 June 1769, it became a separate colony. The first governor arrived in 1770 and immediately appointed a council of seven members; three years later the first Assembly met for the dispatch of business. Subject to the British government's veto, the Island now had the legislative and administrative machinery to manage most of its own affairs. It was renamed Prince Edward Island in 1799 in honour of Edward, Duke of Kent.

From the very outset the Island was bedevilled by the failure of a large majority of the proprietors to honour their two principal obligations: the settlement of their lots in the proportion of one person for every 200 acres (that is, on the average one hundred persons per 20,000-acre lot), and the payment of their annual 'quit rents' (land taxes) varying from two to six shillings per one hundred acres. In 1779, twelve years after the granting of the lands, forty-nine of the sixty-seven townships did not have a single settler, and in 1797, thirty years after the grants, twenty-three lots still had no settlers. Twelve had 216, and six contained 300. In only twenty-six were there the required number of settlers, and the majority of these had not been sponsored by the proprietors. The attitude of the proprietors to the payment of the quit rents was also largely one of non-fulfilment.

For nearly one hundred years the Islanders insisted, in legislative addresses, memorials, and through delegations to London, that since the proprietors had ignored the conditions attached to their land grants, their property should be reinvested in the Crown and regranted to the tenants who tilled the land. The proprietors, however, influential in London, were able to persuade the Colonial Office and parliamentarians that property belonging to them by right of inheritance or purchase and to which anachronistic conditions were attached must be protected.

From the start, these struggles against a distant and seemingly uncaring administration melded the Islanders into a strongly cohesive people, and as the population grew through strong infusions of Irish, Scottish, and English immigrants, the united voices of the Islanders demanding reform grew more strident. By the 1840s the Islanders concluded that their problems could be solved by the acquisition of Responsible Government, which would give them almost complete control of their own affairs. For some years the British government, advised by men such as Sir Donald Campbell, Lieutenant-Governor of the Island, refused to concede such a privilege. 'Four-fifths of the Island constituency', Sir Donald insisted, are 'ignorant, sprung from the pauper classes of the British Isles. From East Point to West Point, there are not twenty men capable of intelligent participation in Government'. The British listened sceptically for a few years, but gradually the intelligent and sophisticated approach adopted by Island politicians such as George Coles and Edward Whelan convinced them that Campbell's charges were grossly unfair, and that Prince Edward Island was indeed ready for Responsible Government. In April 1851 George Coles headed the first responsible government ministry.

It was an auspicious time for the implementation of self-government. Self-sufficiency in food supply, excellent markets for agricultural and fisheries products, and a diversified economy brought prosperity to the Island. The population, which numbered 62,678 in 1848, increased to 80,857 by 1861. Laws providing for free education, reciprocity with the United States, a broadened franchise, and the purchase of the proprietary lands by the government as they became available were placed on the statute books. Although the 'Land Question' was still unsolved, the Island government remained optimistic that compulsory legislation for the sale of the estates would eventually be authorized by the British government. The closely knit Islanders were at the zenith of their political, economic, and cultural independence.

The question of Confederation was introduced into this happy milieu in the 1860s. The position adopted by the Island in the Confederation negotiations illustrates in large measure the traditional characteristics of its politics and its attitude to distant administrations. This attitude might best be described simply as a policy of aloofness. The Islanders, with so profound a respect for self-government, such a high degree of self-sufficiency, and such ardent provincialism, viewed with distaste and suspicion the Confederation scheme. Such a change would of necessity result in an alteration of the Island's constitution, include the Island in a distasteful Canadian tariff structure, and endanger its individuality by placing it under the jurisdiction of a central government in which it would have little influence. Edward Whelan spoke for most of his fellow citizens when he stated that Islanders believed that 'under one large Parliament, legislating hundreds of miles away, their wants would not be as well cared for as they would be by a Parliament sitting within a day's drive of all of them, and directly under the control of all.'

Despite playing host to the Charlottetown Conference in 1864 and taking a vigorous part both in that meeting and in the Quebec Conference one month later, the Island decided to withdraw from the Confederation movement, embarking on a policy of 'splendid isolation'. But in 1871 the Island enacted railway legislation that created a severe economic crisis. Two years later, the economy, close to bankruptcy, prompted the government's reluctant conclusion that Confederation was the only feasible solution to its problem. The Island thereupon entered into negotiations with the Canadian government. The assumption of ownership, maintenance, and operation of a 200-mile Island railway, a guarantee of 'efficient steam service' and 'continuous communication' with the mainland, a loan not exceeding $800,000 for the compulsory extinction of all proprietary titles, and an allotment of six members in the House of Commons were the most significant of the generous terms under which PEI became the seventh province of the Dominion of Canada on 1 July 1873.

PEI's entrance into union with Canada has not lessened its desire to maintain its own political, economic, and cultural identity. Although the basic elements of this identity—the family farm, the independent fishery, small business, closely knit family life, the active church, the small community—are under constant pressures, the maintenance of a meaningful political identity within the federal framework has always been the major problem. During the Island Confederation debates of 1865 on the Quebec Resolutions, the principal objection of Island politicians was the diminution in significance of the Island's Legislature without any real compensating gain in influence in the federal House of Commons. 'In this House', George Coles reminded the Assembly, 'scarcely anything would be left us to do but to legislate about dog taxes and the running at large of swine.' And James McLaren aroused real fears in the Legislative Council when he conjectured that 'with readjustments in representation every ten years, we would probably lose one of our representatives, then another and another, till in the course of time we might lose them all.'

McLaren's words were prophetic. Federal redistribution lowered Island representation to five in 1892, four in 1904, and finally, in 1911, only three. The Island bargained vigorously with the Canadian government, and an amendment was made to the British North America Act whereby no province would have fewer members in the House of Commons than it had senators. Prince Edward Island was thus guaranteed a minimum of four members. This compromise, and others reached through the years, is some recognition that factors other than size, population, and material resources must always be considered in a federal state.

Shortly after PEI entered confederation Lord Dufferin, the Governor General of Canada, paid a formal visit to the Island. During his tour he wrote to Sir John A. Macdonald that he 'found the Island in a high state of jubilation, and quite under the impression that it is the Dominion that has been annexed to Prince Edward Island, and in alluding to the subject, I have adopted the same tone.' Lord Dufferin was a discerning person. Prince Edward Islanders are comfortable as Canadians and love their country deeply, but they also unabashedly treasure their distinctive Island identity, and always appreciate its recognition by others.

FRANCIS W. P. BOLGER
Stanley Bridge, PEI, January 1984

Irish Moss, Skinner's Pond

When white-capped waters roar on the northwestern coast of the Island, its shores and beaches are frequently strewn with a valuable seaweed cut loose from the rocks by the sea's rage. This seaweed, called Irish Moss (*chondrus crispus*), from which carrageen is extracted, is an important cash crop on the Island. The name carrageen is derived from the village of Carragheen, near Waterford, Ireland, where the value of this marine plant was recognized many years ago. Appropriately, the Marine Plants Experimental Station in Miminegash is only a few miles from Waterford, but this one is in PEI. Carrageen is a versatile extract with a wide variety of uses in human food production and industrial application. It possesses emulsifying, stabilizing, and suspending properties that make it useful in the manufacture of more than a thousand products, from ice cream, chocolate milk, salad dressings, and sherbets, to cosmetics, toothpaste, insect sprays, and automobile tires.

There is considerable drama associated with the gathering of moss. Very often the harvesting is a family affair, with women and children flocking to the beaches with the men after heavy weather has sent the waves rolling in with fronds torn from the bottom. While the moss on the shore is being gathered up by rakes and forks, horses with scoops harnessed behind them trot or canter—often to the urging of children—along the ebb waters to salvage the moss floating out of reach. Stompin' Tom Connors, in his 'Song of the Irish Moss', captures much of this excitement when he writes:

> *There's horses in the water*
> *And horses on the road*
> *And here comes old Russell Aylward*
> *And he's hauling up another big load.*
> *And the party lines keep ringing*
> *And the word keeps passing on*
> *You can hear them roar from the Tignish Shore*
> *There's moss in Skinner's Pond.*
> *On old Prince Edward Island*
> *There's one big hullabaloo*
> *The boys and the girls and the old folks*
> *They're going to make a few bucks too.*
> *Getting wet to the neck in the ocean*
> *Where the waves all turn and toss*
> *But it's a free-for-all, and they're having a ball*
> *They're bringing in the Irish Moss.*

Red Road, Long River

Much of PEI's natural beauty is due to vivid colour contrasts—the rich red of the roads, the deep green of the fields, and the brilliant blue of the encircling ocean. The unique characteristic is, of course, the red roads. In the musical *Anne of Green Gables*, the romantic and nature-loving Anne immediately asks Matthew why the Island roads are so red. Matthew hesitatingly suggests: 'Well now...I remember a fellow telling me once it was the iron in the soil getting rusty. But I don't think he could have meant it.' Matthew need not have been sceptical! The rich, red colour is indeed caused by the presence of iron oxide, or rust, in Island soils. Although dust and mud may occasionally lessen the charm, there is nothing more distinctive or more beautiful on Prince Edward Island than the remaining 1,200 miles of red roads that have been spared burial under tons and tons of insipid cement.

Victoria

Victoria, unlike so many Island villages that sprang up haphazardly at rural crossroads, was a carefully planned village by the sea. In the mid-nineteenth century, Donald Palmer, son of the distinguished Island politician James Bardin Palmer and heir to the Palmer estate, laid the foundations for the handsome village. Recognizing the potential of the natural harbour on the fringes of his property, he laid off a corner of his large acreage into blocks, which were then subdivided, leased, and later sold.

Victoria, with its excellent harbour and rich agricultural hinterland, soon became a thriving seaport. Within a short time, its bustling waterfront and busy hotels, stores, and service businesses made it the most important communication and transportation centre between Charlottetown and Summerside. But the same causes that accounted for its growth were also responsible for its decline. As shipbuilding and shipping trickled to a halt and transportation moved increasingly by land rather than by sea, Victoria was gradually bypassed. In recent years, however, the village has enjoyed a resurgence. The beautiful homes and hotels and the places of business are being restored tastefully and finding new purpose as inns, antique shops, restaurants, and artists' studios. The community hall, built in 1914, 'with its slanting floor sloping towards the stage, so there's not a bad seat in the house', features excellent summer stock theatre. A sojourn in Victoria is a passport to an affluent, self-sufficient village of another age.

Sand Dunes, National Park

The high, peaked hills of sand in the national park, partially covered with coarse grass, are among the most fragile of nature's creations. They are appropriately called sand dunes, for they begin as nothing more than small piles of sand, delivered by the sea and shaped by the wind. Determined to survive, these piles of sand are preserved by marram grass, a tough plant that thrives in the barren sand dunes. Its extensive and tangled root system reaches deep into the sand and quietly stills the movement of the wandering sand. It is a mutually beneficial relationship. The marram grass needs the support of the sand to continue its upward growth, and the dune continues to grow as the marram grass holds the sand. If the marram grass is allowed to stabilize the dunes and if there is augmentation by more wind-blown sand, an impressive dune formation will appear. Frequently bayberry, wild rose, beach pea, and lichens will grow on the back slope, also helping consolidate the dunes. The frequent appearance of ripple marks on the surface of the dunes adds to their beauty. The sand dunes patiently continue their centuries-long struggle for survival, now faced with the treasonable destructiveness of man's dune buggies, dirt-bikes, and careless trampling.

Cavendish Sandspit and the Piping Plover

Cavendish Sandspit, which forms the western end of Cavendish Beach, is a paradise for those who love to stroll along quiet, sandy beaches with miles of impressive sand dunes in the background and limitless ocean in the foreground. From mid-May to mid-July, however, such people are asked to leave about five kilometres of their sanctuary for the exclusive use of the piping plover. This small, handsome, secretive shore-bird, whose plaintive whistling calls suggested its name, is an endangered species. Approximately twenty pairs, representing one per cent of the world's rapidly dwindling population, nest in the national park, arriving from the southern Atlantic states in early spring and, like most Island visitors, immediately rushing to the North Shore.

The favourite nesting sites of the plovers are the flat wash-over areas between the dunes and the ocean along Cavendish Sandspit, where an overlay of pebbles, gravel, and broken sea shells provides excellent camouflage for adults, eggs, and fledglings. However, since plovers and their nests are so well camouflaged, people walking along the shore can unconsciously trample eggs and the young plovers. In early August, the piping plovers in increased numbers leave the Island and wing southward to their winter havens and the strolling season returns to Cavendish Sandspit.

Keir's Wharf, Malpeque

Malpeque Oysters (right)

In Malpeque and adjoining districts, Keir is a household name. Rev. John Keir arrived in Malpeque—formerly called Princetown—as a Presbyterian missionary in 1808, the beginning of fifty years of distinguished and zealous service in the area. His son, William, and grandson, James, were both medical doctors who skilfully attended to the medical needs of the people for an aggregate of nearly one hundred years.

John Keir's spacious home, still extant, was situated on Wharf Road on the east side of Malpeque Bay. In the 1840s Keir used the family home as a theological school, and it also served as the site for his sons' medical practices.

The wharf below the house, fittingly enough, is called Keir's Wharf and was once, like the wharves of so many seaside villages, a hub of activity. At its zenith, at the turn of the century, dozens of schooners, loading and unloading produce, berthed at Keir's Wharf and many of the hundreds of oyster boats, farming and fishing the world-famous Malpeque oysters, used its facilities. The Micmac Indians from Lennox Island frequently visited here to barter their crafts for other commodities. Nor was recreation precluded: the deep water off the wharf made it a favourite of swimmers and divers. Today Keir's Wharf—even in its abandoned state—is a silent, yet powerful, witness to the economic and social life of another era.

Prince Edward Island's Malpeque oysters have long been recognized as the finest available to the oyster connoisseur. They are harvested from their deep beds around the Island and have a much better flavour, taste, appearance, and preservative quality than competing varieties. The luscious Malpeque oyster is especially popular for the half-shell trade. The oyster belongs to a large class of mollusks, called Bivalvia, referring to its double shells, or valves. The oyster's two shells are hinged at one end and held together by a central muscle that enables it to open for feeding on the plankton in the sea water. The valves can be closed quickly and tightly, thereby affording protection against predators. The shells are heavy, chalky white, and vary in size and shape according to the bottom conditions.

The average adult oyster spawns about 500-million eggs into the sea water. After fertilization the eggs develop into tiny ciliated larvae, which swim about for two to three weeks. Those that survive settle to a hard surface, after which they are referred to as 'spat'. Oysters take from three to five years to develop from 'spat' to the market size of three inches. Although 'dredging' and handpicking are popular methods of harvesting, the dory, the oyster tongs, and the tonging board are the hallmarks of the typical public oyster fisherman. For seafood lovers, Malpeque oysters are a delicacy and a synonym for succulence.

Drying Codfish, Tignish

The Shipbuilding Industry

Atlantic cod (*Gadus Morhua*) is one of the ground fish intensively fished by Prince Edward Island fishermen. It is caught by otter trawls, line trawls, pair seines, gill nets, jiggers, and various other methods. Cod is marketed fresh, frozen, smoked, canned, or salted. The Tignish Fisheries Co-operative handles some three million pounds of cod annually, 25 to 30 per cent of which is prepared for the boneless salt codfish market.

The proper drying of cod is a highly professional art in which the Tignish Fisheries Co-operative takes justifiable pride. At their ultra-modern plants at Tignish Run and Judes Point, the co-operative is equipped to take care of the heavy landings of headed and gutted cod. After passing through the various stages—icing, splitting, deboning, salting, and pickling—the cod is normally dried in the open air on flakes. These flakes are constructed of wood, and the cod to be dried is placed on frames covered with nylon netting. Since the best grade of fish is produced by sunshine and low temperatures, most of the cod is dried on the flakes in early fall. Cod caught by shore fishermen and 'light' salted can usually be dried in one day. After drying, the boneless salt cod is boxed or packaged to meet the demands of an ever-increasing consumer market.

Apart from agriculture, fisheries, and their related service industries, shipbuilding was the largest employer of labour and the greatest producer of wealth on the Island in the mid-to-late nineteenth century. Great Britain, after her supply of Baltic timber was jeopardized by the Napoleonic Wars, looked to her North American colonies for the timber so vital to the continuation of the war. The next logical step was the building of the ships close to the source of the timber, and timber being one resource the Island possessed in considerable quantity and quality, the PEI shipbuilding industry was born. British entrepreneurs, such as Cambridge, Chanter, Ellis, Hill, Yeo, and Pope, pioneered and developed the industry. As the years passed, such names as Owen, Peake, Orr, LePage, Duncan, MacKay, MacGill, Douse, Bell, Lefurgey, MacDonald, Heard, Welsh, and numerous others were added to the shipbuilding roster.

It may be exaggerating to say that sailing ships were built on every harbour, bay, sheltered cove, inlet, and river of the Island, but it is difficult to name one today that cannot make such a claim. Indeed, a key feature of the wooden shipbuilding industry was its decentralization. Inland from the bays and rivers, men throughout the length and breadth of the Island were kept busy

Hutt Brothers, Boat Building

Shipbuilding Interpretation, Green Park

felling the trees and hauling them to the shipyards; and at the shipbuilding sites others were employed in sawing and hewing lumber, planking and timbering the hulls, fashioning blocks and pulleys, tempering the iron, splicing ropes, caulking seams, sewing sails, and performing all the other tasks connected with the building of schooners, brigantines, brigs, barquentines, and barques.

While the tiny Island's accomplishments throughout the century were commendable, the decade from 1861 to 1870 was the most productive period, 914 vessels being constructed in some sixty yards. The outstanding productivity, however, was all too short-lived. The declining availability of suitable timber on the Island and the advent of iron and steam gradually led to the virtual disappearance of the wooden shipbuilding industry by the turn of the century. All that remains today are some rapidly vanishing traces of the shipyards, the imposing homes of some of the builders, and a rich heritage based on the gigantic accomplishments of another era.

A sense of the romance associated with the building of wooden vessels can be captured by a visit to Hutt Brothers, Alberton, where Gary and Roger Hutt continue the boat-building business, begun in 1962 by their father, Delmont, and his brother, Harvey. Hutt Brothers builds from ten to twelve forty-five-foot fishing boats for Island fishermen each year, in addition to repairing countless others. Island woods—juniper for the timber, black spruce for the planking, and rock maple for the keels— are used exclusively in fashioning these boats. Although Hutt Brothers cater primarily to Island fishermen, they also fill contracts for the provincial and federal governments. The boat in the stocks is being built for walrus hunting and will be manned and operated by the Eskimos in Canada's North. As one watches the men at Hutt Brothers plank and timber the hulls, under the expert guidance of Roger and Gary Hutt, it is apparent that first-class boat building is by no means a lost art in the Maritimes.

Once an active shipbuilding site, Green Park today graphically commemorates the shipbuilding industry of the nineteenth century. An interpretative centre details the construction methods used in building the wooden ships, the history of the shipbuilding industry on Prince Edward Island, and the important role played by the Yeo family in this productive industry. Within com-

Wild Flowers, Stanley Bridge

fortable walking distance, on the edge of Campbell's Creek, is a re-created shipyard that captures the atmosphere of the 1870s. A partially framed vessel with a seventy-five-foot keel, representative in size and design of the type constructed in the 1870s, is the focal point of the shipyard display. All the timber on the ship has been hand-hewn using the same methods employed more than a century ago. Also at the shipyard are a carpenter shop, a blacksmith shop, a steam-box used for bending the ship's planking, and two saw pits. One is tempted—in the mind's eye—to look over one's shoulder to observe a Yeo ship glide gracefully into Campbell's Creek.

To farmers, perennials and biennials such as daisies, dandelions, vetch, foxglove, dropwort, Queen Anne's lace and goldenrod are pestiferous weeds; to many others they are wild flowers. While the former apply herbicides to eradicate them, the latter excitedly focus their cameras to capture their beauty. In the midst of clashing value judgments, the biblical reference seems timeless: 'Think of the flowers; they never have to spin or weave; yet I assure you, not even Solomon in all his glory was arrayed like one of these.'

Lady's Slipper, Floral Emblem of PEI (left)

The pink, or stemless, lady's slipper (*cypripedium acaule*), the most common lady's slipper on the Island and in many ways the most striking of its flowers, has been the official emblem of the province since 1965. This unusual and attractive flower is a member of the exclusive orchid family. Since the lady's slipper is stemless, its two large oval leaves arise from the ground. In early June, a solitary bloom can be found at the top of each of the naked flower stalks, which vary from six to twelve inches in height. These stalks shoot up from the ground between the leaves. One of the three petals of the flower expands into an inflated pendant sac, or 'slipper', which is pale lavender-pink, crimson, dark purplish-pink, or sometimes even pure white. The lady's slipper is commonly found in mixed woodland and in moist places where there is an abundance of rich leaf mould. Like many Islanders, the lady's slipper thrives only in its own habitat and, if transplanted, will normally survive for a relatively short time.

Orby Head, National Park

Although beaches weighed significantly in the selection of Cavendish–Dalvay as the Island's national park, the area is not without many additional contrasting features. The rugged red cliffs at Orby Head rise fifty to one hundred feet from the boulder-strewn shores immediately below, in stark contrast to the sandy beaches. For thousands of years, water and ice have sculpted the Cavendish coastline. From Orby Head, where the bedrock meets the sea directly, the red sandstone on which the province rests is clearly visible. The redness is caused by the iron that binds the sand grains together. Above the sandstone is a layer of glacial till, varying in depth, topped by a thin layer of reddish or brownish-red soil in which grass and occasional low bush and tree cover are rooted. The iron in the sandstone bedrock is weak, and erosion occurs at an average rate of one metre a year. But Islanders are consoled by the fact that much of the eroded material reappears down the coast as a sandy beach.

Kildare Capes (left)

The cliffs and promontories at Kildare Capes are among the most impressive on the Island. Wave erosion has sculpted the coastline into a spectacular series of irregular cliffs of rock and glacial overburden. This seascape is an arresting reminder that PEI is an island indeed, and always at the mercy of the powerful ocean.

Lobster Trap Setting Day

The laying of the lines, signalling the opening of another lobster season, is an exhilarating occasion. Family, friends, and neighbours customarily gather at the best vantage points in the stillness of the early morning to watch the drama unfold. The months of preparation, with painting of boats, fine tuning of motors, building of lobster traps, and preparation of the warp and plastic buoys, rapidly fade into oblivion as the lobster fishermen rev up their motors and begin manoeuvring from dock-side. At a given signal by a fisheries department official, hundreds of boats, heavily laden with baited lobster traps, speed out to sea to lay claim to their favourite fishing grounds, and trap after trap is quickly lowered to the ocean floor.

The Blessing of the Boats, North Rustico (left)

The blessing of the boats in North Rustico and many other fishing communities is an impressive ceremony with important religious and social connotations. The fishermen and their families, their boats decorated with flags and bunting, sail past the wharf as the local clergyman prays that 'God will bless the fishermen and their boats, and that he will protect the fishermen and give them a bountiful catch'. The ceremony and the festive mood it engenders inspire a wonderful sense of togetherness for the many families who depend for their livelihood on the generosity of the sea.

The Wharf, Stanley Bridge

Stanley Bridge, a small, picturesque village, is situated at the junction of Stanley River and New London Bay. Called Fyfe's Ferry until 1865, Stanley Bridge enjoyed the self-sufficiency and prosperity that characterized so many Island villages from the mid-nineteenth to mid-twentieth centuries. Located in a prosperous agricultural area and enjoying all the advantages associated with excellent water communication, Stanley Bridge became a thriving commercial centre. The wharves, built where the bridge crosses Stanley River, served the farmers and merchants who moved most of their produce and merchandise by schooners. Today, in addition to fulfilling its traditional role as the home port of its fishermen, Stanley Bridge Wharf is the site of a lobster pound, oyster operation, and deep-sea fishing business.

Lobster Traps, New London

Before and immediately after the lobster season, Island wharves are neatly stacked with the traps that the fishermen use to catch the delectable lobster. They have been unchanged in design since the turn of the century, with the exception that laminated bows, nylon nets, and polypropylene warps usually replace fir bows, cotton, Manila, and cedar. Lobsters enter a trap to consume the bait in the first compartment. The ring and netting are slanted to allow entry and prevent exit. Once inside, they eat some of the bait and then look for a way out. A short trip through a second ring or funnel leads them into prison and seals their doom. The popularity of the lobster has made the lobster trap a treasured souvenir. Traps leave the Island by the thousands on the rooftops of tourists' cars to become conversation pieces in homes across the land.

The PEI Lobster

The fishing industry has traditionally played an important though fluctuating role in the Island's economy. The industry's seasonal nature and the belief that a reasonable subsistence was more likely to be assured from mother earth rather than from the restless and unpredictable ocean led the vast majority of Islanders to opt for agriculture and other professions. Within recent years, however, government subsidies and earned unemployment insurance benefits have resulted in more and more Islanders, making fishing their exclusive career.

Lobster fishing, pursued during a two-month season, is the mainstay of the industry. Today, 1,301 licensed lobster fishermen, each usually employing one helper, fish with 377,650 traps from some seventy Island ports in the Gulf of St. Lawrence and Northumberland Strait. The lobster, *Homarus americanus*, is a decapod crustacean and is usually greenish-blue speckled with dark spots when alive and a bright red colour when boiled. Lobsters normally inhabit rocky bottoms and feed principally on fish, other crustaceans, and shellfish. The lobster's large

North Lake

claws are effective in capturing food and warding off enemies. If the lobster is suddenly disturbed, it escapes by flexing its powerful tail muscle and shooting backward.

Since the lobster is such a high-priced epicurean dish, it is difficult to believe that at the turn of the century they were so plentiful that they were used as land fertilizer. Indeed, lobster at one point was the trademark of the poor; many school children, in order to avoid the scorn of their classmates, pleaded with their mothers to make their sandwiches with jam, jelly, or molasses rather than the ridiculed lobster meat. Today the lobster is the undisputed monarch of the sea and the gustatory delight of millions. The lobster remains mysterious and elusive, but lobster fishermen with bigger and more powerful boats, replete with echo-sounders and modern navigational instruments and the latest in efficient traps, hydraulic trap haulers, and handling gear, are successful enough in meeting the demands of an ever-increasing market.

Eptek Centre, Summerside

Eptek Centre, situated on Summerside's waterfront, is a national exhibition centre and the home of the PEI Sports Hall of Fame. The centre was built with funds supplied by the 1973 PEI Centennial Commission and the National Museums of Canada. Because of its national mandate, Eptek Centre features some excellent travelling exhibitions from the National Museums in Ottawa and from other Canadian museums. In addition, some major and minor local exhibitions are mounted each year. The name of the centre is singularly appropriate. The early Micmac Indians, who camped along the shores of Bedeque ('Bedek') Bay, called the area 'Eptek', which means 'hot spot' or 'hot place'. The sculptured cedar doors symbolize this fascinating native place-name. The mandala and sun motifs, hand carved in relief, speak of the 'sunny camping place' on which this popular cultural centre is situated.

St. Dunstan's Basilica, Charlottetown (right)

One of the grandest edifices on the Island is St. Dunstan's Basilica, whose magnificent spires dominate the skyline of Charlottetown. This graceful structure is the fourth that has been built on its site. The first chapel, built in 1816, was named St. Dunstan's at the request of Bishop J.O. Plessis of Québec, in gratitude for the kindness and attention he received from the English civil authorities during his episcopal visit to the Island in 1812.

Churches built in 1843 and 1897 preceded the present building. St. Dunstan's Basilica, 271 feet in length and 90 feet in width, was designed by J.M. Hunter. Five years in construction, it was dedicated in 1919. It is built in the form of a Gothic cross, with its Gothic twin spires on the east front rising to a height of 200 feet. The church is constructed of Wallace and Miramichi stone. The interior, with its terrazzo floor, marble columns, and embossed vaulted ceilings, is truly magnificent. Above the 37-foot marble altar is a beautiful rose window, which was fashioned in Munich, Germany. Notable features of this basilica are the stained-glass windows in the façade. Designed by Henry Purdy, they depict St. Dunstan, the renowned Archbishop of Canterbury, Angus Bernard MacEachern, the first bishop of Charlottetown Diocese, Pope John XXIII, and Pope Paul VI, with the Glorious Resurrection in the centre.

Wood Islands ferry

Tuna Fishing (left)

Bluefin tuna, a large food and game fish of the mackerel family, ranging in weight from 500 to 1,500 pounds, has been fished off the shores of Prince Edward Island for some fifteen years. Since the industry on the Island began at North Lake, and world records are made and broken there on a regular basis, this bustling port justifiably bills itself the 'Tuna Capital of the World'. The world record for a Bluefin tuna, 1,496 pounds, is currently held by Ken Fraser of North Lake.

All tuna caught in the Gulf of St. Lawrence must be caught with rod and reel or with tended line. Sports fishermen, fishing usually on chartered boats, must use rod and reel with a 130-pound test line. The normal practice is to attract the tuna with a wire leader baited with several mackerel. The end mackerel has a hook embedded in its stomach, while the remaining six act as 'teasers'; as this baited leader is trawled behind the boat, it gives a school-of-mackerel effect.

The tuna fisherman is harnessed into a swivelling 'fighting chair' bolted to the deck and is expected to keep the fastened rod and reel at the ready. If a torpedo-like Bluefin strikes, the fisherman in the chair matches wits, strength, and endurance with a cunning monster, yielding line when necessary and reeling in to gain line as the fish tires. It can take from a fraction of an hour to ten hours or more to bring the trophy fish to gaff. If the fisherman wins the struggle, the charter returns to port with the flag of success floating from the mast. At the port, the photographers and the buyer take over. The successful fisherman, armed with substantiating photos, has enough stories to last a lifetime; the owner of the boat has the money realized from the sale and the tail of the tuna to nail to his fishing shack; and the consumers in Japan who buy the fish will soon have a favourite delicacy.

Tobacco Farming

Tobacco production on PEI was started experimentally by the provincial government in 1959. Since then, tobacco growing has become increasingly popular; today some seventy tobacco farmers grow approximately 4,000 acres of the plant. Tobacco is a hardy plant with a coarse, central stem, terminated at the height of six to eight feet by a long flowering stalk. Fifteen to eighteen leaves grow in whorls of three from this central stem and vary somewhat in size, shape, and texture, depending on their position on the plant. The plant prefers a well-drained sandy-loam soil, and PEI soils are, therefore, ideal. It also needs a warm, sunny growing season with a minimum frost-free period of 120 days. The possibility of early frost poses the greatest hazard for the tobacco farmers on the Island. From early August until the end of September, tobacco workers 'prime' the man-sized tobacco plants in the fields, while in the kilns the owners attend to the fine art of curing the tobacco. Steady markets for carefully cured leaves have made the growing of tobacco a highly productive operation. The potato still reigns supreme on the Island, but tobacco is on the march and making a bid to dethrone the mighty spud.

The PEI Potato

One of the most distinctive features of the Prince Edward Island landscape is the regular pattern of immaculately cultivated potato fields on farm after farm in the picturesque, rolling countryside. PEI is appropriately called 'Spud Island', having earned the title in the production of what is recognized the world over as the finest stock of seed and table potatoes. Seed potato certification began in 1916, when a few strains of Irish Cobbler and Green Mountain were found to be free of viral diseases.

From these beginnings, the Island has developed and refined the quality of its vigorous, disease-free seed potatoes to the degree that, today, they are sold to dozens of countries and account for 60 per cent of Canada's annual export of this agricultural commodity. Although processing potatoes on the Island is a relatively new operation, it is rapidly becoming more important. In addition, the Island's flavourful, high-quality table potatoes, recognizable in Canada's major food markets by the premium label and by traces of characteristic red Island soil, have long been a favourite of consumers. Approximately 800 Island farmers normally plant a total of 71,000 acres of

Potato planting, Kingston

potatoes, 75 per cent of which are certified for seed with the balance grown for table-stock and processing. Kennebecs, Sabagoes, Netted Gems, Superiors, Red Pontiacs, Keswicks, Russet Burbanks, Cherokees, Katahdins, Green Mountains, and Irish Cobblers are the most popular varieties. The heavy investment in land, machinery, fertilizers, and storage facilities and the almost incredible risk regarding price make potato growers the most heroic people in the Island's agricultural industry. From late April to early June, the potato fields are alive with feverish activity as the potato farmers engage in planting operations made urgent by the fact that a growing period from 100 to 140 days is required for potatoes to mature properly. The planting phase of the operation is facilitated by the use of modern two-to-six-row planters, which make the planting drudgery of another era a faint memory. After the crop is planted, the potato farmer must frequently cultivate the drills and apply insecticides and sprays to prevent blight. Farmers drive up and down that same row with their machinery up to fifteen

Sandy MacNeil on the family farm, Clyde River

times during the course of the four-month season, making the potato field their summer home. Top- or vine-killing time is late August or early September, and harvesting begins about two weeks later. The harvesting and subsequent grading operations vividly illustrate the phenomenal mechanization of the potato industry in recent years. Two methods of potato digging are generally used on the Island. Both employ mechanical harvesters, as opposed to the older diggers that required individual bagging in the field, and both commonly use bulk trucks with self-unloading hoppers to transport the tubers to ultra-modern storage facilities. The first method employs a two-to-six-row harvester to dig the potatoes directly. The second employs a wind-rower, which deposits the potatoes it digs in one row and is followed by a harvester taking up the tubers, as in the first method. Some growers sell directly from the field, but the majority store their crop in warehouses, later to be graded and sold.

The Blue Heron

The Blue Heron (*Ardea herodias*), the tallest bird on the Island, is a familiar figure on the edges of bays, rivers, ditches, mud flats, marshes, dunes, lakes, and ponds. Alert and motionless, the stately bird relies on lightning-quick jabs of its snake-like neck to capture the fish it so cautiously hunts. If disturbed at its work, it leaps effortlessly to the air, croaking indignantly, and flying away with legs trailing. This fine-feathered fisherman is the symbol of the Prince Edward Island National Park.

Haslam's Inn, Springfield

Among the most commodious houses of the early nineteenth century were the 'half-way houses' or inns that abounded on the Island. In the year 1825 there were twenty-six road-houses in Charlottetown alone and forty-three registered elsewhere in Prince, Queen's, and King's counties. In the early days of the Island, these houses existed primarily to provide lodging, food, and beverages to their clientele. Nevertheless, they often assumed nobler and loftier purposes. The Cross Keys Tavern in Charlottetown, for example, was the setting for the first session of the PEI Assembly, in July 1773. One may speculate that the environment may have been at least partially responsible for doorkeeper Edward Ryan's irreverent observation that it was a 'damn queer parliament'. When Bishop J.O. Plessis of Québec visited Charlottetown in 1812, he found comfortable accommodations at Samuel Bagnall's Inn. With no Catholic church available in the city, he celebrated Mass in the parlour of McPhee's Tavern on Dorchester Street.

The 1841 minutes of the presbytery of the Presbyterian Church in Prince Edward Island contain an account of a meeting 'held at Haslam's Inn', a large white house situated in Springfield built by Thomas Haslam in 1834. He built this twelve-roomed 'half-way house' to take advantage of the new post road that had recently been built between Charlottetown and Princetown. Like the proprietors of all licensed road-houses of the period, the Haslams were required to 'keep three good and sufficient feather beds and bedding for the accommodation of travellers with good stalled stabling and necessary and wholesome provender for six horses'.

Haslam's Inn, situated on a lovely eminence, still stands today in excellent condition, the oldest home in Springfield. It is owned by Thomas Haslam's great-grandson, Basil, his son John and John's wife Jean. The stately rooms, the floor-length windows, the spacious hallways, the elegant nineteenth-century furnishings, and the architectural niceties of the exterior combine to make it a gracious reminder of the past.

One-Room School, Orwell Corner

The agricultural historic site at Orwell Corner attempts to re-create a typical turn-of-the century rural crossroads. Pivotal to this interpretation is the one-room schoolhouse located on the Belfast Road. Built in 1895, it was used until school consolidation in the 1960s. The desks and other furnishings admirably capture the atmosphere and flavour of the 'seats of learning' of another era.

The Rural School

To today's school children, chauffeured from their homes in comfortable buses to spacious and elaborately equipped regional schools, the one- and two-room schoolhouses of two decades ago seem almost like historic relics, as, perhaps, they did even to those who attended them. One of the few constants in Canadian society for the past few decades has been school consolidation. In PEI this process, although comparatively slow in arriving, has been relentlessly pursued. Before consolidation, the basic pattern for the organization of schools was the small rural school-district system. Some 475 school districts of approximately five square miles were established, with the schools centrally located and at least three miles apart.

One- and two-room schools (and the occasional three-room school) were thus built across the Island. The government paid the basic salaries of the teachers, but the district was responsible for the construction of the school and payment of the teacher's supplement. Each school, administered by a locally elected school board of three members, became a solidly entrenched rural institution of central significance in the maintenance of a strong Island community. Since each district enjoyed a high degree of financial and administrative autonomy, the quality of the school building and facilities and the qualifications of the teacher often bore a direct relationship to the priority the district placed on education, with well-appointed schools in one district contrasted with primitive buildings in the adjoining district. While the controversy continues as to the advantages and disadvantages of consolidation, there is no gainsaying the applicability to rural PEI of the following lines from Oliver Goldsmith's *Deserted Village*:

> *Beside yon straggling fence that skirts the way,*
> *With blossomed furze unprofitably gay,*
> *There, in his noisy mansion, skilled to rule,*
> *The village master taught his busy school; ...*
> *While words of learned length, and thundering sound,*
> *Amazed the gazing rustics ranged around,*
> *And still they gazed, and still the wonder grew,*
> *That one small head, could carry all he knew.*
> *But past is all his fame. The very spot*
> *Where many a time he triumphed, is forgot.*

Wild Lupines, Stanley Bridge

Wild lupines provide some of the grandest flowers on the Island. Their blue, purple, white, yellow, red, or pink pea-like blossoms are tightly set along one foot or more of their erect two-to-five-foot stems. Their handsome foliage is deeply divided into many palm-like segments. Lupines thrive in sandy soil and propagate generously from seeds sown in gardens or scattered with reckless abandon along the roadsides. From mid-spring to midsummer the long, bright clusters of wild lupines painting the ditches and hillsides add greatly to the charm of the Island.

Kensington Railway Station

The Legislature's authorization in April 1871 of the construction of a railway between Alberton and Georgetown was the first stage on the Island's ride into joining Confederation, which the province was forced to do when the high cost of building the line added substantially to the Island's public debt. The contractors, with government approval, built a serpentine and community-minded 147-mile railway to service the actual 120-mile distance between Alberton and Georgetown. One of the areas accommodated by the circuitous extension of the line was Kensington. Peter Sinclair, a legislative representative from the district, made a revealing remark when he was chided with responsibility for the Kensington extension: 'I am accused of being the cause of making one curve in the line. I did use my influence to bring the road to Kensington, but I did not ask to have it turned right back again to Bedeque!'

In any case, the town of Kensington is now graced with a striking station, constructed in an era when railway stations were intended to be architecturally pleasing. The handsome exterior, comprising walls made of small granite boulders embedded in cement, is complemented by a lovely interior with walls and wainscoting in pine and ash and hardwood floors throughout. Formally opened in 1905, this station was designated a national historic site in 1978.

Yeo House, Green Park

James Yeo, jun. (1827–1903), 'hunchback Jemmy' as he was commonly known, was a member of one of the leading shipbuilding families on the Island. His father, James Yeo, sen. (1789–1868), was perhaps the most colourful of a long list of shipbuilder–merchant magnates of the period. His son inherited much of his father's business acumen and established his own shipyard at Campbell's Creek, a short distance from the elder Yeo's operation at Port Hill. Like his father, he was a politician, and he enjoyed a distinguished career in provincial and federal politics.

But there the resemblance seemingly ends. James, jun., was a cultivated gentleman and did not share his father's unbridled ruthlessness. Moreover, the refinement and peacefulness of his enterprise at Campbell's Creek is in sharp contrast with the stark ruggedness of

Spring day, Charlottetown

his father's establishment at the busy crossroads at Port Hill. James, jun., chose a beautiful site for his shipyard, where, between 1856 and 1886, he launched some twenty-three ships, some of them the largest built by the Yeo family.

In 1865 he built a handsome house and called the place Green Park. The exterior of the house is solidly Victorian. Noteworthy is the octagonal cupola that surmounts the main roof. From the floor of this cupola, approximately six feet above the ceiling of the third floor, James Yeo was able to observe the whole Yeo enterprise at Green Park and at the ancestral home at Port Hill. He lived in the stately house in Green Park until his death in 1903. Green Park itself is now a provincial historic park, and James Yeo's house, largely unaltered, has been tastefully restored.

36

City Hall, Charlottetown (left)

Charlottetown city hall's formal beginnings were ushered in with unusual panoply. On 1 July 1887 the city concluded its two-day celebration of Queen Victoria's Diamond Jubilee with an impressive parade through its principal streets to the intersection of Queen and Kent. There the participants proudly watched the laying of the cornerstone of the new city hall by the Hon. John Yeo, Grand Master of the Masonic Order.

Although Charlottetown had been incorporated in 1855, a converted court-house and a market-house served as city halls until 1888, when the present commodious brick structure was built. Designed by Charlottetown architects Phillips and Chappell in Romanesque Revival style, the tower and round arches are the most striking features of the building. City hall has provided space for the Council Chamber, the mayor and clerk's offices, Stipendiary Magistrate's Court, the police station, the fire department, and various other municipal offices.

The saga of 'Big Donald' (a bronze fire bell that replaced the town crier in the 1870s, named after Donald MacKinnon, chief of the fire department in 1875), is intimately associated with the building. 'Big Donald' was first located in the tower of the old wooden market building. In spite of two trips back to Boston for recasting, 'Big Donald' served the city well and in 1888 was moved to the tower of the new city hall. It continued to sound fire alarms until it was made redundant by the installation of a modern air horn. In 1966 'Big Donald' was tenderly removed from the tower and now rests in appropriate dignity in front of city hall.

The Farmers' Bank of Rustico

In 1859 the largely Acadian community of South Rustico became the beneficiary of the talents of an enterprising and innovative individual, Georges-Antoine Belcourt, parish priest. Among his early accomplishments were the establishment of a high school in his own house, in which he also taught, the founding of *L'Institut de Rustico* for the promotion of adult education, and the acquisition of an annual grant from Emperor Napoleon III of France for the endowment of a library.

But Father Belcourt is best remembered for his economic initiatives. He believed that a sensitive lending institution, in which farmers could procure loans for agricultural purposes at reasonable rates of interest, was needed to improve the minimal living standards of his people. Thus was born a farmers' bank. To house the bank, Father Belcourt designed and engineered the construction of a large stone building. The red island sandstone, quarried in Hope River, was brought to Rustico in sleighs as his parishioners made their sacrificial Lenten treks to the church.

Begun in 1861, the building was ready for occupation when the act for incorporation of the bank received royal assent, in April 1864. The functional two-storey building served the farmers and fishermen of the area until a compulsory system of large-scale nation-wide branch banking led to the expiration of its charter in 1894. The Farmers' Bank of Rustico, formally designated a national historic site in 1971, was the smallest bank ever to operate in Canada and was a precursor of the credit union movement.

Ardgowan National Historic Park, Parkdale

Ardgowan (hill of the daisy) was the home of William Henry Pope, one of the Island's Fathers of Confederation. Pope, the eldest son of another prominent politician, Joseph Pope, was born at Bedeque in 1825. He received his early education on the Island and then went to England for higher studies, reading law at the Inner Temple, London. As a lawyer, land agent, editor of *The Islander*, politician, and judge, Pope was intimately involved in the public affairs of his time.

When the Pope family lived at Ardgowan from 1854 to 1873, it was a gracious residence with an estate of some seventy acres. W.H. Pope, affable and gregarious, loved to entertain, often to the depletion of the family finances. During the Charlottetown Conference in 1864, Pope hosted the delegates at Ardgowan and provided what George Brown described appreciatively as 'a grand *dejeuner à la fourchette*, oysters, lobsters, & champagne and other Island luxuries'. After the Island's entry into Confederation, Pope was appointed a judge of the Prince County court, an acknowledgment of his legal expertise and of his consistent and dedicated advocacy of Confederation.

Ardgowan remained in the Pope family until 1879. Through the years it became almost a relic of its former self: the beautiful house and gardens were allowed to deteriorate, and the estate was reduced to less than five acres. Fortunately in 1967 Ardgowan was acquired by Parks Canada to commemorate the Island Fathers of Confederation, among them Pope. The district office of Parks Canada is now located in the carefully renovated house, and the grounds have been restored to the horticultural fashions of the nineteenth century. In 1982 Ardgowan was formally designated a national historic park.

Dalvay-By-The-Sea

As PEI moves through the last quarter of the twentieth century, an increasing number of imposing summer and retirement homes bedeck its shoreline. It seems unlikely, however, that many will match palatial Dalvay-by-the-Sea, a splendid example of the large summer homes erected by wealthy citizens in an age when land and building costs were comparatively low and income taxes were non-existent!

In the 1890s Alexander MacDonald of Cincinnati, a wealthy director of Standard Oil and an early partner of John D. Rockefeller, purchased a 160-acre tract of land in the Tracadie area and built Dalvay House, named after his birthplace in Scotland. Situated on a placid lake, within breathing distance of the Gulf of St. Lawrence, Dalvay-by-the-Sea is a multi-gabled, two-storey structure of wood and Island stone, embraced by a massive veranda and a *porte-cochère*. The entrance hall has a baronial ambience, with a two-storey-high foyer and a massive Island sandstone fireplace that comfortably consumes three-foot logs.

For some fifteen years the MacDonald family spent their summers on the Island, entertaining regally. After MacDonald's death in 1910, and a subsequent decline in family fortunes, Dalvay-by-the-Sea was sold to the caretaker. It passed through various hands until its purchase by the national park in 1936. Parks Canada now takes impeccable care of the whole estate and leases Dalvay-by-the-Sea for operation as a summer hotel.

Town Hall, Summerside (right)

Shipbuilding, excellent harbour facilities, merchandising, small manufacturing, the fox industry, and a rich agricultural hinterland enabled Summerside to evolve into a sophisticated urban centre. One of its most historic buildings is the town hall. This distinctive building, constructed and used as the post office until 1952, had a tentative beginning. Jean MacFadyen, in *For the Sake of the Record*, describes the circumstance: 'An unusual sidelight is the fact that this building was designed and intended to face in the direction opposite to which it was constructed. When the foundation had been completed, it was discovered that an error had been made in the reading of the federal blueprints and the only way of overcoming that mistake was to redesign the front.'

Built of brick, this imposing Victorian structure was completed in 1886, at the modest cost of $31,000. The installation of a four-sided illuminated clock, in 1915, added to the building's distinctiveness. This venerable structure, centrally situated in the town, has served as Summerside's town hall since 1956.

41

Province House, Charlottetown

The three-storey Georgian Province House sits proudly in the centre of Queen Square, Charlottetown, the unmistakable symbol of PEI's self-respect. At the opening of the Legislature in 1837, Lt.-Gov. John Harvey observed that the Island needed a 'colonial building' to accommodate its public records. The Island Assembly enlarged on his suggestion and the building eventually included administrative offices, the two Houses of the Legislature, and the law courts.

A design competition was held throughout the Maritimes, and the £20 prize was won by Island architect Isaac Smith, who had already achieved considerable distinction through his designs of Government House, the Central Academy, and other public buildings on the Island. The cornerstone was laid in May 1843 by Lt.-Gov. Henry Vere Huntley, and in January 1847 the Legislature met in the colonial building for the first time.

Excellent craftsmanship characterizes this finely proportioned stone building, with its dominant porticos and classic-revival detailing. Fittingly enough, Province House was built and furnished by Islanders; only one of the eight construction contracts, and that for the supplying of Nova Scotia Wallace sandstone, went to a non-Islander. Furniture was made by such prominent Island cabinet-makers as Charles Dogherty and Mark Butcher.

Architect Smith designed the building to accommodate the principal administrative offices and the Supreme Court on the first floor; the library and conference room at the centre of the second floor, with the House of Assembly and Legislative Council at opposite ends; the galleries overlooking the two houses; and additional offices on the third floor.

Prince Edward Island is justly proud of its venerable Province House, where Responsible Government was won, where the first Canadian conference on Confederation was held in September 1864, and where its House of Assembly still meets. Formally designated a national historic site on 1 July 1983, Province House ranks as one of Canada's most important historic buildings.

Confederation Chamber, Province House, Charlottetown

In this historic chamber, between 1 September and 7 September 1864, twenty-three delegates—eight from Canada, five from Nova Scotia, five from New Brunswick, and five from Prince Edward Island—reached agreement on the desirability of a federal union and on the general outline of a future constitution. Perhaps most importantly, they dedicated themselves to the building of a new nation. The Legislative Council Chamber, now the Confederation Chamber, provided a gracious setting for this famous Conference. The chamber, tastefully restored by Parks Canada in the early 1980s, captures the 1864 ambience described by Mary K. Cullen in her book on Province House. 'The spacious chamber spanned the width of the building and rose the height of two storeys to a graceful arched ceiling adorned with a plaster oval centerpiece and plaster panel moldings and corner ornaments. Fluted ionic columns supported a narrow gallery on three sides of the room which was entered from the third floor. An oak grained rail and banisters enclosed the balcony while a similar balustrade separated the entrance from the Chamber proper.'

Since the six days of closed meetings at the Conference had produced such excellent results, it seemed appropriate that the concluding banquet and ball should be held in Province House. *The Examiner* described the affair as 'the most brilliant Fête that has ever occurred in Charlottetown'. The Lieutenant-Governor's office on the first floor served as a robing room, and the Legislative Council Chamber was transformed from a conference centre to a reception and drawing-room. The Legislative Library—always crowded—was converted into a refreshment room with abundant quantities of tea, coffee, sherry, and champagne. The Legislative Assembly room, where the guests danced to music supplied by two bands, was decorated with mirrors, flags, evergreens, and flowers. To climax the occasion, the Supreme Court room was converted into a banquet hall, where guests found the tables 'literally groaning under the choicest viands prepared in Mr. Murphy's choicest style'. Can anyone wonder after all this that Prince Edward Island is the 'Cradle of Confederation'?

Government House, Charlottetown

Through the early years of colonial government in PEI, the official representatives of the British Crown were obliged to find their own accommodations. Some bought or built houses, others were billeted with the military, and still others rented residences. After sixty years of such haphazard arrangements, the Island government decided that an official residence was needed for the Lieutenant-Governor. The site had been chosen in 1789 when Lt.-Gov. Fanning designated a beautiful hundred-acre tract overlooking Charlottetown Harbour as the location of a future residence for the representative of the Crown.

At Fanning Bank in 1833–4, local contractors Isaac Smith, Henry Smith, and Nathan Wright constructed a finely proportioned frame structure of neo-classic design. The most notable visual feature of the residence is a portico supported by four pillars that extend the full height of the house. The attractiveness of the exterior is complemented by the spaciousness and striking formality of the interior, which is highlighted by the wide and lofty entrance hall with Doric columns and a grand staircase illuminated by a Palladian window.

Before the attainment of Responsible Government in 1851, Government House served as the seat of executive power. Through the years this house has provided a setting of simple elegance for the entertainment of the Island community and its distinguished visitors. It has been the host to royalty, governors-general, and other notable personages, but perhaps most proudly of all to the Fathers of Confederation who attended the Charlottetown Conference in 1864. During that historic conference the Fathers were fêted at a dinner party and at a formal ball, the latter event described facetiously by George Brown as 'a very nice affair, but a great bore for old fellows like me'. The residence was also the formal setting for a rare and memorable photograph of the Fathers. Designated as a national historic site in 1973, the venerable 150-year-old Government House will continue to play a significant role in the unfolding of PEI history.

'Beaconsfield', Charlottetown

James Peake, a prosperous Charlottetown merchant and shipbuilder, built this elegant house in 1877. He employed a young Charlottetown architect, William Critchlow Harris, to design the house and spent lavishly of the Peake fortune in its construction. In *Gothic Dreams*, Harris's biographer, R.C. Tuck, describes this fine example of Victorian architecture: 'Built of wood in Second Empire Style, it had 25 rooms, a mansard roof, topped by a belvedere, and was fronted on its western side by a veranda. Inside, high moulded ceilings, richly furnished hardwood doors, encaustic tile, marble baseboards, and a Romanesque staircase window with James Peake's initials in stained glass all contributed to an atmosphere of luxury.'

The Peakes had enjoyed their luxurious residence for only six years when financial disaster dissipated the family business. Beaconsfield was offered for sale by the mortgagor, William Cundall. It was perhaps some consolation in later years that James Peake's young son could always excite the envy of his fellow Islanders by relating the story of how he and a future king (George V) sneaked into the Beaconsfield wine cellar during a formal reception and got 'properly drunk'.

When no purchasers for Beaconsfield appeared, the Cundall family moved in and lived there until 1917. The Cundall will directed that Beaconsfield should serve as a Christian residence for young women. Managed judicially by the Cundall Trust, the house served respectively for some fifty years as a home for working girls, for girls attending Prince of Wales College, and as a residence for nurses of the Prince Edward Island Hospital. In 1972 Beaconsfield became the home of the newly established Prince Edward Island Heritage Foundation, and on 3 July 1973, tastefully refurbished, the residence was officially opened by Her Majesty Queen Elizabeth II.

46

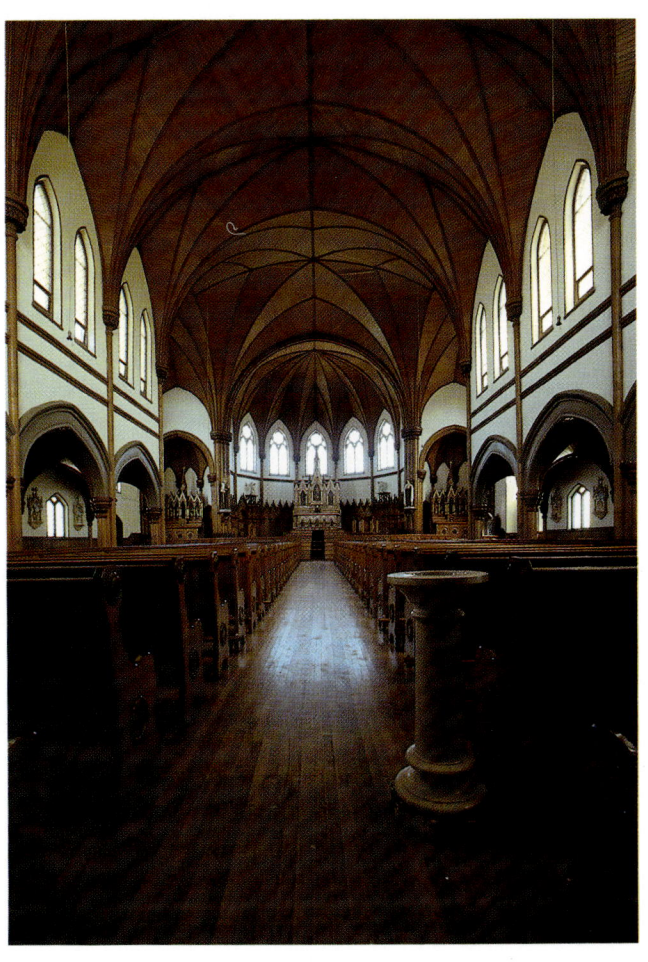

St. Mary's Church, Indian River

The first St. Mary's Church was struck by lightning and burned in 1896. The incumbent, Mgr. D.J. Gillis, is reputed to have declared, when he saw the impressive St. Malachy's Church in Kinkora: 'Yes, yes, the new St. Mary's must be built like Kinkora, only bigger and better.'

His first move towards this objective was to engage the same architect, William Critchlow Harris, to design the church. St. Mary's Church, completed in 1902, at the moderate cost of $20,000, is one of the most beautiful wooden churches on the Island. Built in Gothic style, its most striking exterior feature is its round spire with a circlet of arched niches at its base occupied by statues of the twelve Apostles. The interior of the church is awe-inspiring. Island woods, notably spruce, maple, fir, and birch, are used almost exclusively for the furnishings, which, specially designed by Harris, afford to the interior, as R.C. Tuck remarks, a remarkable 'unity and harmony of style'. The groined roof of lightly stained spruce gives a lofty and ethereal effect. Unfortunately the pulpit, designed by Harris to be partially lowered into the floor by a system of pulleys, was not built. One can only speculate on the dramatic conclusion of sermons with pulpit and priest disappearing from sight!

The Hon. George Coles Building, Charlottetown (left)

Standing next to Province House, the former Law Courts Building, now designated the Hon. George Coles Building, was erected in 1874-6, with Thomas Alley as architect. The original Law Courts Building of pressed brick, with iron cresting surmounting its mansard roof, and featuring a large illuminated clock in the tower of its south façade, was severely damaged by a disastrous fire in 1976. Fortunately the building has been tastefully renovated and now serves as a legislative annex and home of the provincial archives. The new name gives appropriate recognition to a distinguished Island politician. The father of Responsible Government, free education, a broadened franchise, and land reform, the Hon. George Coles, was also a Father of Confederation and a premier of the province.

Church of Scotland, DeSable

Among the most appealing features of Island landscape are the white wooden churches dotting the countryside and contributing immeasurably to the Island's pastoral charm and serenity. The Church of Scotland at DeSable, on a quiet wooded hill overlooking the DeSable River and Northumberland Strait, is representative of these architectural gems. This imposing structure was built in 1855 by the Rev. Donald McDonald, who ranks as one of the most dominant spiritual figures of the nineteenth century. During a forty-year ministry on the Island, McDonald single-handedly created and ministered to a religious body of some 5,000 adherents.

While theoretically McDonald remained a Church of Scotland minister, the members of his congregation became known as 'McDonaldites'. One of the spiritual phenomena associated with the McDonald ministry was the manifestation on the part of many during his services of extraordinary excitement characterized by unusual distortions of body and limbs. David E. Weale, an authority on McDonald, writing in *The Island Magazine*, places these spiritual experiences in their proper perspective: 'These strange physical manifestations of piety were referred to as "the work" or "the works," and were to become an accepted part of McDonald's ministry throughout his career. In fact, in subsequent years McDonald's followers were often referred to derisively as the "kickers" or the "jumpers", and it is unfortunate that while "these works" were only one aspect of the "McDonaldites" faith, they became, in the popular mind, the customary means of identifying the group.'

The large white church at DeSable, scene of many memorable services conducted by McDonald, reflects the Scottish tradition in its architecture, especially in the high pointed windows, wrought-iron ornamentation in the spire, and finely tapered steeple topped with a thistle.

Ice Racing, North River

Ice racing on the bays and rivers of Prince Edward Island has always been a popular winter pastime. In the Charlottetown area, the focal point for winter horse racing for many years was the harbor ice opposite Fort Edward in Victoria Park. The races, attracting Island-wide competition and fans, were of a quarter of a mile distance with the times normally ranging from thirty to thirty-five seconds. In the 1950s, the ice adjacent to the North River Causeway replaced Charlottetown Harbour as the locale for winter racing. The introduction of pari-mutuel betting has added to the popularity of the sport. When ice conditions and weather permit the staging of races, large crowds gather on the ice and wonder perhaps whether Ike Moreside's world-record drive with Hal McKinney in twenty-eight and one-fifth seconds will be equalled or surpassed.

50

Provincial Exhibition and Old Home Week

The precursor of the provincial exhibition and of the several agricultural fairs held throughout the province was staged at Crapaud in 1820. The advertisement for this fair simply stated that 'the inhabitants of Crapaud and surrounding county are anxious to open a market with Ramshag [now Wallace, Nova Scotia], to exchange sheep for spinning wheels and chairs.' It also announced that 'young cattle of the best breeds from Cumberland, Nova Scotia, would meet with purchasers for cash or barter.' Fairs were not pretentious in the beginning, but they helped improve the quality of agricultural crops and conformation of livestock. Their broader appeal was sparked in 1888 by the incorporation of the Charlottetown Driving Park and Provincial Exhibition Association. Two years later the association established a complex featuring an elegant exhibition building and a raceway on a forty-acre site overlooking the Hillsborough River. The first combined horse-race meet and provincial exhibition was held there in October 1890.

Annually since that time livestock of all kinds, agricultural produce, horticultural products, foods and preserves, and handcrafts of every description have been on competitive display in the exhibition building. While the horse races have always been an important ingredient of the provincial exhibition, they gradually assumed a more dominant role, and in the late 1940s the first Old Home Week was inaugurated. Initially during Old Home Week, the Charlottetown Driving Park sponsored ten-dash harness-racing programs for seven afternoons and evenings. But with the establishment of the prestigious Gold Cup and Saucer event and the lengthening of Old Home Week to ten days with the addition of Country Days, PEI's midsummer 'madness' has reached dizzy heights matched only by the ferris wheel in the adjacent midway.

Atlantic Wind Test Site, North Cape (left)

Complementing the lighthouse at North Cape is the unique 185-acre Atlantic Wind Test Site. Created by the governments of Canada and Prince Edward Island, the test site is designed to facilitate research, development, testing, and demonstration of wind-energy technology and equipment in Canada. The site was established and developed through the ingenuity of the PEI Institute of Man and Resources and carries out its research in co-operation with the National Research Council of Canada. The test site also offers and operates an independent laboratory service to industry, governments, businesses, and other agencies that require technical and scientific assistance in the vital and growing wind-energy field.

The facilities are located in a harsh but typical coastal environment in the Gulf of St. Lawrence, with an average wind speed at the site of 14 miles per hour. It seems fitting that PEI, which made considerable use of wind as a source of energy in an earlier era, is providing expertise for experimentation in this all-important alternate energy field.

Harness Racing, Charlottetown

Harness racing has always been a popular sport on the Island. From early spring (year-round in Summerside and Charlottetown), trotters and pacers hold the spotlight, which is not surprising, since Islanders have always been incorrigible horse lovers. The horse was the principal means of transport until the 1930s, and nearly every household had a 'driving horse'. Many were the challenges as to horse supremacy issued and settled on the way to church and school and even as funeral processions solemnly wound their way down the road or across the winter ice.

The Summerside race track had already staged the famous Hernando-Black Pilot Race in 1888 and more than a dozen other ovals were scattered throughout the province when the Charlottetown Driving Park held its first race in October 1889. Some of the popular rural tracks have been at Northam, New Annan, Upton, Souris, St. Peters, Pinette, Murray Harbour, Hope River, Cymbria, Covehead, Montague, and Alberton (Joe O'Brien's home track). It was Milligan & Morrison's magnif-

Mare and foal, York Point

icent oval at Northam in 1932 that first introduced Islanders to night racing under the lights. The day of the races at these rural tracks almost took on the status of legal holidays. The thousands in attendance could enjoy not only the races but also musical entertainment, boxing bouts, excellent meals, games of chance, and more than occasional unscheduled pugilistic encounters.

Today two-minute miles, *pari-mutuel* betting, and starting gates are striking symbols of the sport's sophistication. Many veterans will insist, however, that it is impossible to surpass the colour and drama of the old scores; the fans heaping abuse on the drivers manoeuvring for advantageous positions before they got the word 'Go' from the stentorian starter; the rush down the tracks in high-wheeled sulkies, the drivers placing the butt of their whips between the big wheel spokes and creating a noise that seemed to coax extra effort from the horses and to increase the excitement of the fans.

54

Milligan and Morrison's barn, Birch Hill (left)

Hunter River

When the mixed family operation was the way of life for rural Islanders, the farmhouse was usually given subordinate status; indeed, it was the rare family that did not spend considerable money improving the barn while denying themselves modern conveniences in the house. The barns, large and small, that enrich the countryside have a majesty that speaks for itself. The Milligan and Morrison's barn at Birch Hill, built in 1925, was at the time the largest and most modernly equipped in the Maritimes. At the barn's opening fête, attended by some 7,000 people, speeches, meals, cattle judging, and track and field events were climaxed, *The Agriculturist* noted, 'by a grand dance in the capacious loft gorgeously decorated with cedars, myriads of electric light bulbs, Japanese lanterns, and umbrellas'. This barn has cathedral-like proportions and an ingenious arrangement of timbers forming the main loft and mows. And what can surpass the beauty and lure of the red barn? Even the Island's barns that are in retirement and decay are merely exposing slowly the secrets of their builders and owners.

Bonshaw

Modern steel barns with their striking silos rising like so many steeples are gradually dominating the landscape, and more and more Islanders are left with childhood memories of *their* barns. But these wonderful memories—of cows and calves, horses and foals, seasoned wood, the smell of crops, games in the hay mows, and swallows circling the rafters, ridgepoles, and beams—are enough to last a lifetime.

Marshfield

Indian Point Lighthouse

The Indian Point lighthouse, constructed in 1881, is located on the south side of the entrance to Summerside Harbour. Perhaps the most unusual of all PEI lighthouses, Indian Point consists of a forty-two-foot octagonal tower, rising from the centre of the dwelling that stands on a circular pier, which has its foundation below the low-water mark. A breakwater was constructed in 1885 to protect the lighthouse. Today this uniquely shaped harbour light, now called Indian Head sector, with a range of eighteen miles, still commands the busy traffic lanes leading in and out of the harbour.

Blockhouse Point Lighthouse (right)

The lighthouse at Blockhouse Point is one of the most architecturally pleasing in the system. Situated on the south west side of Charlottetown Harbour, Blockhouse Point has always enjoyed a strategic location. During the latter half of the nineteenth century, as freight and passenger streamers began to make weekly voyages to and from Charlottetown, a modern guidance system for Charlottetown Harbour became imperative. James W. Butcher of Charlottetown was awarded a contract to construct a tower and dwelling house at Blockhouse Point. Completed in 1877, this handsome building, enjoying a beautiful setting, has been a mecca for visitors through the years. Virtually unchanged since 1877, it is a pleasant reminder of the excellent workmanship of PEI's nineteenth-century craftsmen.

59

East Point Lighthouse

As early as the 1850s petitions for the construction of a lighthouse at East Point were presented to the PEI Legislature. Despite its obvious necessity for the safety of the substantial commercial traffic at the most easterly point of the Island, it was not constructed until 1867. Dudley Witney, in *The Lighthouse*, notes the East Point lighthouse's unique distinction: 'It is the most peripatetic building of its kind in the country.' A local contractor, William McDonald, built the sixty-four-foot-high tower about a half-mile inland from East Point. Octagonal in shape, it was handsomely girded with hand-hewn timber and sheathed with shingles.

In 1882 the British warship *Phoenix* was wrecked offshore, and the tragedy was at least partially blamed on the fact that the lighthouse was too far removed from the coast to be clearly visible in bad weather. As a result, in 1885 the lighthouse and keeper's cottage were moved closer to the coast. At the same time a fog-horn alarm building was constructed and two fog horns installed.

In 1908 the foundations were found to be eroding. The lighthouse was once again jacked up and moved, this time 200 feet inland to its present location.

This wooden lighthouse is a worthy representative of the practical craftsmanship of the nineteenth century. In the interior, rugged beams and knee-joints blend harmoniously with sculptured banisters and ornate mouldings. Wonderfully preserved and one of the few in PEI still manned, this lighthouse stands proudly as the Island's most easterly sentinel, contemplating, perhaps, its next move.

North Cape Lighthouse (right)

The geography of PEI and the extremely dangerous reef at North Cape were ample justification for the establishment of a lighthouse at the most northerly point of the Island. But it took some twenty-five years of relentless petitioning before the Legislature issued a contract for the structure. In 1865 this sixty-two-foot lighthouse tower was built and, in the following year, a keeper's cottage was built close to it. Now fully automated, the lighthouse at North Cape is of invaluable assistance to those who ply their trades in some of the best fishing and Irish Moss grounds off the coast of the Island.

61

62

Point Prim Lighthouse (left)

Preparing for races, Summerside Harbour

In spite of the Island's inviting appearance, its treacherous reefs, narrow harbours, and dangerous capes and cliffs often present serious navigational challenges to those who travel by sea. During the nineteenth century, marine traffic around the coast of the Island increased with the expansion of commercial shipping and the fishing industry, making necessary the installation of lighthouses and other sophisticated range lights.

The first lighthouse to grace the verdant shores of the Island was built at Point Prim in 1845. The Island's political authorities decided to give special attention to such a major navigational inauguration. Isaac Smith, the Island's leading architect, with a design for Province House already in hand, was commissioned to draft the plans for the lighthouse and keeper's cottage at Point Prim, and the site was chosen by a legislative committee headed by Joseph Pope, Speaker of the House of Assembly.

The Islander vividly described the occasion: 'On Monday last, a Committee appointed by the House of Assembly...proceeded in ten sleighs to Point Prim. On landing, a site was chosen for the building, which commands a beautiful view of some thirty miles on the Straits of Northumberland, the different points at that distance being easily distinguished. The party partook of a lunch and returned the sixteen miles, in one hour and twenty minutes, thus showing the facility with which travelling can be performed on good ice in winter.'

Smith designed a sixty-foot circular brick tower, topped by a polygonal lantern, and a separate wooden keeper's cottage. The lighthouse was lighted for mariners for the first time in December 1845, although Smith's architectural plans were not completely implemented until 1847 when the brick exterior was boarded in and shingled. In a wonderful state of preservation, this handsome pharos still continues its faithful vigil that began some 150 years ago.

The Yankee Gale

One of the worst natural disasters in the history of the Island was the Yankee Gale of 1851. In those days American fishermen, principally from the ports of Portland, Gloucester, and Boston, came by the hundreds to the Gulf of St. Lawrence to exploit the area's valuable fisheries. On Friday, 3 October 1851, an American fleet of more than one hundred under sail had enjoyed good weather and a heavy catch of mackerel in the section of the gulf that lies in the crescent made by the northern concave of the Island. Towards evening an ominous calm was followed shortly by gale-force winds, which left the vessels and their crews in a quandary. Some managed to escape to sea and a few rode out the storm. But the vast majority floundered helplessly in the mountainous waves that accompanied the gale, which lasted two days. When the horrendous storm abated, wreckage strewed the North Shore from East Point to North Cape, and lifeless bodies were found in the holds and cabins, strapped to the rigging and lying on the shore. Conservative estimates placed the number of ships wrecked at eighty, and the number of men lost at 160.

Islanders responded to the disaster with alacrity. The survivors were taken into their homes and given food, clothing, and shelter, and the dead were buried with dignity at Cascumpec, Malpeque, Yankee Hill, Cavendish, Rustico, Tracadie, and other cemeteries on the North Shore of the Island.

Among the pathetic tales associated with this storm is that of the *Franklin Dexter*, which was wrecked on Cavendish Capes. Among the dead were four sons and a nephew of the owner of the vessel, Captain Wickson, who lived in Dennis, Maine. After the storm they were buried in Cavendish cemetery. Later the broken-hearted father came to Cavendish, had the bodies exhumed, and placed their coffins on a ship, the *Seth Hall*, bound for the United States. He then returned home to await their arrival. Captain Hall, who was in charge of the schooner, defied advice and the tides and, before weighing anchor, cursed the storm and its devastation and swore with an oath that he would sail out of the harbour that night, even if he sailed straight into hell!

He *did* sail out of the harbour, but after rounding East Point, the ship and crew were never heard of again. Perhaps the unfortunate captain should be pardoned; the North Shore had been a veritable Hades during those October days of 1851. The oil-on-canvas painting by marine artist George Thresher, dated 1851, is in the permanent collection of the Confederation Centre Art Gallery and Museum.

The Northumberland Strait Crossing

Northumberland Strait, which Islanders are wont to say cuts off the rest of Canada from PEI, has been a decisive factor in determining the province's distinctive identity. The argument of the proprietors that 'the tedious and expensive voyage to Halifax which during the winter months was impracticable on account of the ice' clinched the case for the Island's being made a separate British colony in 1769. Islanders resisted both Maritime and federal Union largely because of their distrust of distant and largely inaccessible administrations and their desire to have, as Edward Whelan aptly expressed it, 'a Parliament sitting within a day's drive of all of them, and directly under the control of all'. And when the Island, in straitened circumstances, reluctantly entered Confederation in 1873, one of the assuaging advantages was the dominion government's pledge to provide efficient steam service for the conveyance of mails and passengers between the Island and the dominion, thereby placing the Island in continuous communication with the Canadian railway system.

Through the years Islanders have interpreted this guarantee with considerable imagination. The 'conveyance of mails and passengers' was construed to embrace the movement of all types of produce and merchandise; 'continuous' was taken to mean daily and even hourly; and 'efficient' the best and most modern communication available. In addition, Islanders insist that 'steam service' should not be interpreted literally but rather should mean any method of connection with the mainland, be it hovercraft over the strait, a tunnel under the strait, or a combination of tunnel and causeway.

Apart from one abortive effort to provide a causeway, the Dominion has attempted to honour its pledge through the provision of ice boats and a series of ships, some satisfactory and others unsatisfactory. These vessels have ranged in efficiency from the *Northern Light* (whose stern was better adapted to breaking ice than her stem and who, therefore, often bumped across the icy strait stern foremost), to the gracious *Abegweit II*, which displays some of the elegance of an ocean liner. Yet the constant comings and goings of these ships across the strait are visible reminders of the Islanders' tenuous connection with the rest of Canada.

66

The Dairy Industry

Dairying has always been one of the more stable types of farming since the arrival of the Acadians. It was not until some forty co-operative cheese and butter factories were established in various Island communities in the 1890s and early 1900s, however, that it became an organized industry. These factories were the heart of the dairy industry until the middle of this century, when centralizing philosophies and a drastic decline in the number of farmers led to the closure of all but half a dozen.

Today the 941 dairy farms in the province can be categorized according to the end uses of the milk they produce. The 508 farms making up the largest group sell industrial milk, used in the manufacture of evaporated milk, cheese, ice cream, and powdered milk, while the 140 making up the smallest group sell only cream, used in the manufacture of butter.

Holsteins are by far the most numerous breed of dairy cattle on the Island, and Ayrshires are also popular. There are fewer farms with dual-purpose Shorthorns, Jersey, and Guernsey cattle, but these are noted for the quality of their stock. In total there are now some 23,500 dairy cattle producing milk for the Island's fourteen processing plants—seven fluid plants, five creameries making butter, and two industrial milk plants.

Although the total number of milk producers has decreased, milk production has been increasing steadily in both quality and quantity, which is reflected in the fact that PEI is an exporting province for all dairy products, while the other Atlantic provinces have deficiencies in dairy production. But dairy self-sufficiency is not easily maintained. In a society becoming increasingly leisure conscious, the seven-day-a-week dairy business is a demanding occupation, one that should not be accepted lightly.

Nowhere is the radical transition that has taken place in agriculture more apparent than in the modern dairy and beef farms. A typical farm homestead of the 1930s comprised a large L-shaped house nestled near an orchard of fruit trees and complemented by a huge barn with many doors, which housed the horses, cattle, and crops, a woodshed, a well house, an outdoor privy, a hen-house, a granary, a chicken house, a goose house, and a storage shed for wagons, sleighs, and machinery.

In today's age of agricultural specialization, the mixed-farming family operation is now largely a memory. Instead, the countryside is dotted with glistening steel barns housing the latest in automatic feeders, cleaners, fans, heaters, not to mention row after row of carefully groomed dairy cattle. Attached to the barn is the nerve centre of the operation, the milk room, housing the huge bulk tank, the modern symbol of sanitation. In close proximity is the towering steel silo or silage barns, filled, or waiting to be filled, with hay and corn silage. A paved driveway to the milk room ensures winter delivery. If we substitute or add loafing barns, bulging at the seams with breeds such as Herefords, Shorthorns, or Aberdeen Angus, we have a modern beef operation. These agricultural changes have dramatically altered the appearance of the countryside.

Eel net, Naufrage   *Eel Fishing, West River* (right)

The eel found in the waters of PEI is known as the American eel, or *Anguilla rostrata*. Its spawning ground is the Sargasso Sea, located in the Atlantic Ocean south of Bermuda and east of Florida and the Bahamas. The larval eels proceed north, passively following the Gulf Stream; after approximately a year—and a journey of some 2,000 miles—many of them make their habitats in Island waters. If the grown eel manages to avoid the spear, the eel trap, the sharp eye and beak of the heron, and other predators, later—perhaps in fifteen years—it will return to the Sargasso Sea to spawn and, presumably, to die.

Eel fishing, a sport for some, provides a reasonable supplement to the income of some three hundred licensed fishermen. On calm nights in early spring and summer, on the bays, rivers, and streams of the Island, the muffled oars of the dories signal the arrival of the eel fishermen, armed with eel spears and bright lights preparing to challenge the elusive eel. Spearing is prohibited during the late summer and autumn trapping season when the bulk of the eels are harvested. When the heavy frost arrives, eels generally burrow into the mud and spend the winter under the ice cover. The fishermen meet with good success spearing the sluggish eel through the ice.

69

West River

Fox Farming

Much of the charm of the Island is due to the network of large and small rivers that journey peacefully through the countryside. Among the most attractive of these is the Elliott or West River stretching some twelve miles from Charlottetown Harbour to the Bonshaw area. Since this river effectively cut off from the rest of the Island the Rocky Point to Bonshaw section of the South Shore, the people of the West River districts responded by making the river their main highway. Even after improved roads and new bridges were built to facilitate and shorten the journey to and from the capital, West River was never ignored. The *S.S. Harland*, which moved people and produce for some twenty-seven years, did not discontinue service until 1935, when faster gasoline-powered launches rendered it obsolete.

But the West River was always much more than a means of summer and winter communication. Its rich bottom surrendered, and continues to surrender, oysters, eels, smelts, and mussels to professional fishermen. And for those who enjoy sport fishing, boating, canoeing, kayaking, or simply basking in beautiful surroundings, the picturesque West River is pure poetry.

In the 1880s Charles Dalton, a farmer, fisherman, trapper, and fur buyer, conceived an idea that brought him wealth, a knighthood, and the office of Lieutenant-Governor, and gave PEI an industry for which, particularly during the first two decades of this century, it achieved world-wide fame: fox farming. Dalton, in the pursuit of his first love—hunting and trapping—had occasionally caught one of the Island's rare black foxes, whose pelts invariably brought the highest prices, and he began to acquire breeding stock in the hopes of developing a strain of pure black foxes from the black pups tipped with silver guard hairs that the native red foxes occasionally dropped.

He met with indifferent success, however, until forming a partnership with Robert T. Oulton, an old fishing and hunting companion. In 1894 the latter built a ranch on Cherry Island (Oulton's Island), which he owned, and Dalton brought the first pair of black foxes.

The two men made an excellent team. As Robert Allan Rankin, writing on pioneer fox farming, has put it: 'Dalton was outgoing and possessed business acumen, while Oulton was a more experienced, innovative and persistent stockman.' Their ranch later served as a model for many who entered the industry. In the early 1900s Dalton, with some misgivings, sold a pair of breeders to Robert Tuplin, who formed a partnership with James Gordon. A further merger with fox breeders B.I. Rayner and Silas Rayner led to the formation of the Big Six Combine, which monopolized the domesticated fox world until 1910, when a Dalton–Oulton sale of twenty-five fox pelts for some $34,000 became public knowledge. The combine came under constant pressure from farmers, merchants, and professional men and was irretrievably broken when the flamboyant nephew of Robert Tuplin, Frank Tuplin, sold breeding stock to anyone willing to pay his almost unbelievable price of $5,000 a pair. There was no shortage of buyers and the golden era of the silver black fox dawned.

A veritable fox mania engulfed the Island from 1910 to 1940, with fabulous prices for foundation stock offered to and accepted by Dalton, Oulton, and the other original fox breeders. Within a few years of the ending of the Big Six Combine monopoly, the Island was literally dotted with fox ranches owned and operated by companies, partnerships, and private individuals. While many ranchers stood pat with the strains perfected by the Big Six Combine, others, such as the famous Milligan and Morrison ranch at Northam, developed their own distinctive strains.

The appeal of fox farming was by no means restricted to the large operators. Small farmers, fishermen, clerks, and labourers got involved and established small ranches on their properties. While World War I caused a decline in the value of the fox pelt, it did not spell a serious disaster. The industry recovered during the 1920s and 1930s and expanded in Canada and the United States. Many lost money, but it is equally true that many were spared severe economic hardship during the Great Depression. While PEI did not enjoy the control over the industry it had exercised in the heady days of Dalton and Oulton, it still held a prestigious position through the Dominion Experimental Fox Ranch at Summerside. The outbreak of World War II, however, marked the end of fox farming as a nation-wide industry through a combination of factors: tariffs, taxes, shortage of feed, material, and labour; foreign competition, governmental regulations; new chemical dyeing techniques; and over-production. An exciting era, in many ways analogous to the gold-rush bonanza, had ended.

After the collapse of the European fur market in the 1940s, only a few die-hard ranchers continued to raise foxes. Today, with the strengthening of the fox-pelt market, the industry is undergoing a remarkable revival. In the changeable and unpredictable world of fashion, fox fur is in once again, especially that of the silver fox and its lighter-coloured mutants, the platinum and the pearl platinum. And Prince Edward Island is once more gaining international prestige for its high-quality silver furs, some from breeding lines developed and refined at the turn of the century. Some 180 registered ranches, with an estimated 4,500 foxes, are involved in the industry.

George Campbell, Park Corner

Although the location of fox ranches varies greatly, established ranchers frequently locate in secluded, wooded habitats. A typical ranch consists of several wire pens complete with dens or nest boxes, surrounded by a high-perimeter 'guard fence'. The den, a box-like structure, is attached to the pen and serves as the 'private quarters' of the fox. It is divided into an outer and inner compartment, the latter providing a comfortable private place for the vixen to whelp her pups.

The fox calendar begins between late January and early March, when one male is mated with several females. After a gestation period of fifty-two days, the mothers whelp, usually bearing an average of four pups each.

After the nursing period the pups are fed throughout the summer and fall on a high-protein diet, including meat, fish, bone meal, skim milk, bread, and some added supplements. By November the foxes have grown their sleek winter hair and are considered 'prime'. After pelting, fleshing, shaping, drying, and cleaning, the furs are ready for market and before the lapse of many months will grace the coat-racks of exclusive stores throughout the world.

Island Fiddling

The importance attached to fiddle music at weddings, house parties, dances, and community concerts gave the fiddler a status in the community rivalled only by that of the doctor and clergyman. The fiddler was always assured of the choicest of food and drink lest he threaten to leave the party ere the break of dawn! In modern times, movies, radio, television, 'name bands', and centralized entertainment have presented formidable challenges to fiddle music. Increasingly, however, Island musicians are 'rosining' their bows and entertaining with their strathspeys, jigs, and reels. Enthusiastic fiddlers' associations have been formed, and phenomenal strides have been taken in fostering the fine art of fiddling.

Joseph G. MacDonald; Sailing Ship Model

When you enter Joseph G. MacDonald's driveway in Borden, you will immediately notice a weathervane in the shape of a sailing ship, just one sign of Joe MacDonald's encyclopaedic knowledge of sailing-ship history and tradition. Transportation in general and sailing ships in particular have been his lifelong fascination, and he says, with only the slightest twinkle in his eye, that 'the most beautiful things in the world are a sailing ship and a pregnant woman.'

MacDonald has been building models of sailing ships as a hobby since 1945. Although he prefers to build ships with three masts, the vast majority of his clients prefer those with two. His love of the sea comes naturally, for Northumberland Strait has always been an intimate part of his life. He was employed with C.N. Marine for thirty-five years as a fireman and oiler, and, in addition, he simultaneously tended two range lights for the Department of Transport. Since his retirement in 1976 he has translated some of his boundless enthusiasm into the establishment of a transportation museum in Borden. He served as curator there for three years and contributed 40 per cent of the artifacts on view from his excellent personal collection. Although MacDonald may not admit it, he is a living custodian of sailing-ship history on Prince Edward Island.

Simpson's Mills, Bay View

The mill at Bay View, operated by Rupert Simpson and his son Herbert, is a segment of living history. A mill has stood in this peaceful valley at the headwaters of Hope River for some 150 years. Indeed the association of five generations of Simpsons with the milling business has resulted in the surrounding area being commonly called Simpson's Mills. In earlier years, the Simpsons operated a carding mill, a grist mill which ground wheat, oats, and barley, and a saw mill. Lumber is still sawed there as is evidenced by the sweet smell of sawdust inside the mill and the stacks of slabs and freshly sawn lumber on the outside. Although power for the operation is now supplied by diesel, the much more romantic water-wheel is still operable. Through the decades, the milling operation has brought the Simpson family into a close relationship with their neighbours. Lumber for many of the houses and barns in the area was, and still is, crafted in this mill. Simpson's Mills proudly stands, austerely practical, supremely efficient, a symbol of stability and permanence in a rapidly changing world.

Cavendish Beach, National Park

The close to eight square miles of North Shore front property, extending along twenty-six miles of attractive coastline from New London Bay on the west to Tracadie Bay on the east, was chosen in 1936 as the site of the Prince Edward Island National Park. The park also comprised the lands in the Cavendish area surrounding the farmhouse known as Green Gables; Rustico or Robinson's Island; a portion of Brackley Point; and an area in the vicinity of Tracadie that incorporated Dalvay-by-the-Sea. The decision of the federal government to locate the park on the North Shore of the Island was not made lightly. Some twenty-two sites were considered, but the present area was chosen primarily for its attractive beaches, central location, unlimited potential for surf bathing, canoeing, boating, deep-sea fishing, and hiking, and for the picturesque drive and sea vista between Cavendish and Dalvay.

The park areas of Tracadie, Dalvay, Stanhope, Brackley, and Rustico had traditionally attracted the wealthy and leisure-conscious to its hotels and beaches; and after the publication of Lucy Maud Montgomery's *Anne of Green Gables* and its sequels, the Cavendish area quickly became an irresistible mecca. Montgomery herself states that Cavendish was chosen as the western section of the national park because it was already a sort of shrine on account of her books and because it had a magnificent sand beach situated between the beautiful New London and Rustico Harbours. And a magnificent beach it is indeed! The glistening white sand beach is as fine as dust and as firm as a billiard ball. With three miles of majestic sand dunes forming an impressive backdrop and with safe, refreshing salt water for swimming, surfing, and boating, Cavendish Beach is a lovely ocean playground.

Exhibition and Festival Acadien, Abram's Village

One of the distinguishing characteristics of PEI society is its closely knit family and community life. Islanders manage to find adequate expression for their cultural, religious, and recreational aspirations within the confines of their communities. Oyster festivals, boat regattas, fishing derbys, lobster carnivals, potato-blossom days, harvest days, plowing matches, and Acadian festivals follow in rapid succession, as one Island community after another celebrates its individuality. One of the most colourful of these events is the Egmont Bay and Mount Carmel Agricultural Exhibition and Festival Acadien.

The agricultural exhibition, now in its eightieth year, features competitive displays of handcrafts, grains and vegetables, cattle, sheep, swine, poultry, rabbits, fruits, and honey. Acadian handcrafts, enjoying an unrivalled reputation on the Island, are eagerly purchased during the exhibition. The Festival Acadien, held simultaneously with the Exhibition, is a lively demonstration of the richness and vitality of Acadian culture. The expressive folk dances, performed by *Les Danseurs Evangéline* and other groups on various occasions during the festival, are unforgettable reminders of the Acadian traditions dating back to the seventeenth century and illustrate how fortunate Islanders are that some two to three hundred Acadians avoided the cruel clutches of Lord Rollo during the 1758 Expulsion.

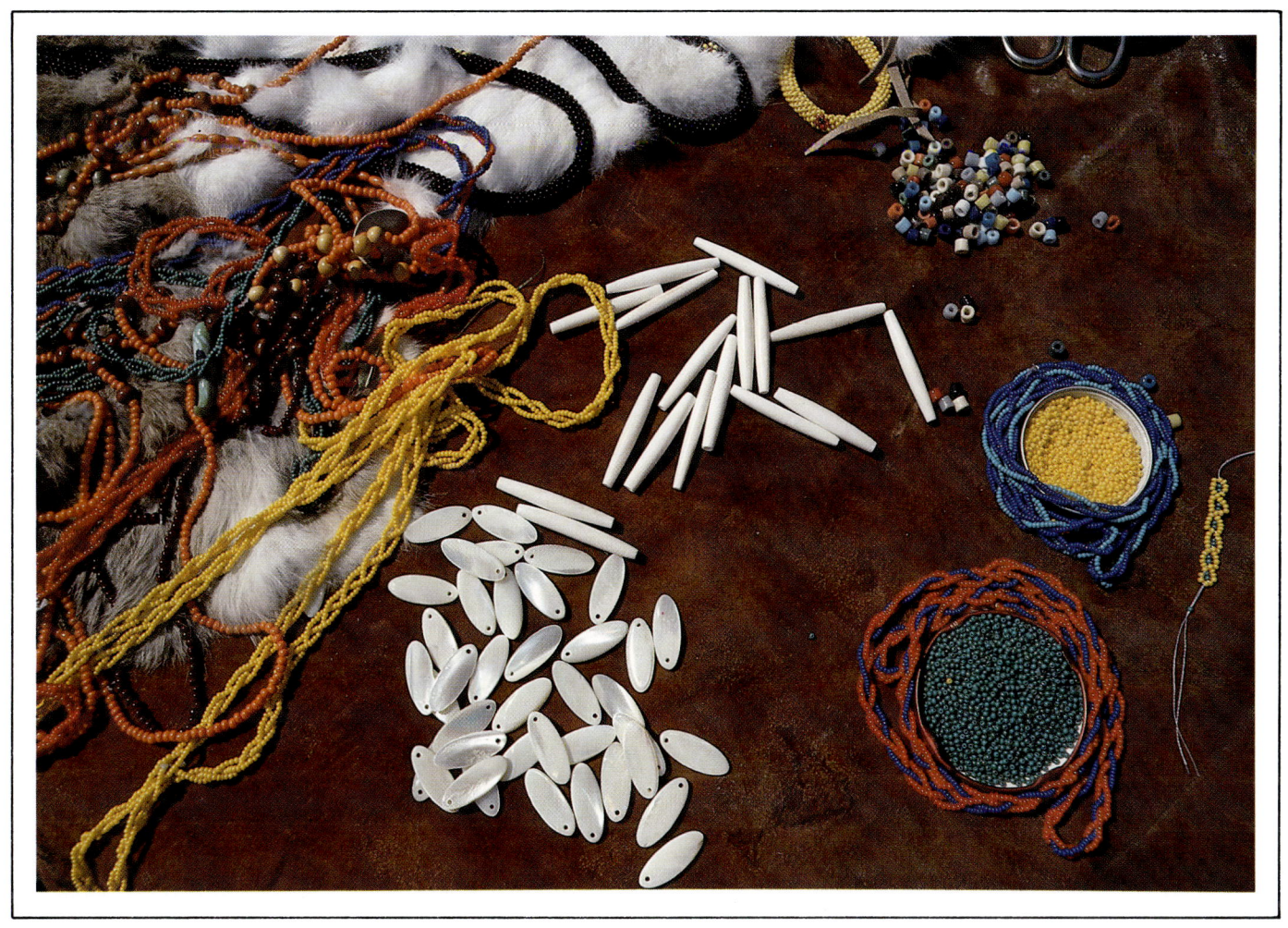

Micmac Indians, Lennox Island

For many centuries the nomadic Micmacs hunted and fished quietly on their beloved Abegweit. All this changed with the movement of Europeans into the area. With the relentless encroachment of white settlement, the Micmacs were gradually relegated to four reserves. The life of the Micmacs altered drastically; they nevertheless retained the high quality of their craftmanship. Today, on the Lennox Island Reservation, for example, there are many illustrations of their quality art. The woven baskets, clay pots, and beaded belts and necklaces are a few of the many examples of the serviceable and decorative Micmac handcrafts.

Confederation Centre of the Arts, Charlottetown

The cultural life of the Island advanced appreciably in 1964 when Confederation Centre of the Arts was opened as a national shrine to the Fathers of Confederation, whose initial meetings had been held in Province House, Charlottetown, in September 1864. The construction costs of the centre, totalling some six million dollars, were met by a grant of fifteen cents per capita by the governments of Canada and of each province, as well as by other sizeable donations. Prince Edward Island provided the site for the complex, adjacent to Province House in Queen's Square. The design chosen by a panel out of forty-seven designs entered in the competition was submitted by the Montreal firm of Affleck, Desbarats, Dimakopoulous, Lebensold, and Sise.

On 2 February 1963, Robert L. Stanfield, premier of Nova Scotia, turned the first sod, and Jean Lesage, premier of Québec, delivered a stirring address. In August of the same year, Prime Minister Lester B. Pearson laid the cornerstone with the ten Canadian premiers in attendance. And on 6 October 1964, Her Majesty Queen Elizabeth II officially opened the centre—Canada's first national centre of the arts.

The stone and concrete complex contains a 1,099-seat theatre, an art gallery and museum, a memorial hall, a 300-seat lecture theatre, and a restaurant, all connected by concourses, foyers, and staircases. The centre-piece of the entire complex is Memorial Hall, a sixty-foot-square structure in stepped stone with a glass roof of dome-shaped pyramids. Behind the glass doors of the hall stands venerable Province House, the *raison d'être* of the whole complex. This magnificent structure, the home of the Robert Harris collection and of the summer festival, which annually features the famous musical *Anne of Green Gables*, plays a vital role in the development of Canadian visual and performing arts.

Anne of Green Gables Musical

The 1965 Charlottetown Summer Festival at the Confederation Centre of the Arts featured a musical based on Lucy Maud Montgomery's classic *Anne of Green Gables*. In 1964, Mavor Moore, artistic director of the Centre, commissioned Don Harron and Norman Campbell to prepare a contemporary musical version of the novel. Their remarkable musical has been an unqualified success. Superbly directed and choreographed by Alan Lund, it has been the anchor production of the Charlottetown Festival each summer and has graced stages and thrilled audiences in London's West End, New York, Osaka, Japan, and various cities of Canada, in four cross-country tours. Now in its twentieth consecutive season, *Anne of Green Gables* is Canada's most widely acclaimed musical of all time. The wide-eyed, exuberant, and romantic Anne Shirley has so captivated audiences that, when the tears recede, they can be heard singing and whistling the theme song and Matthew's final touching tribute to Anne.

> *Anne of Green Gables never change,*
> *I like you just this way.*
> *Anne of Green Gables, sweet and strange,*
> *Stay as you are today.*

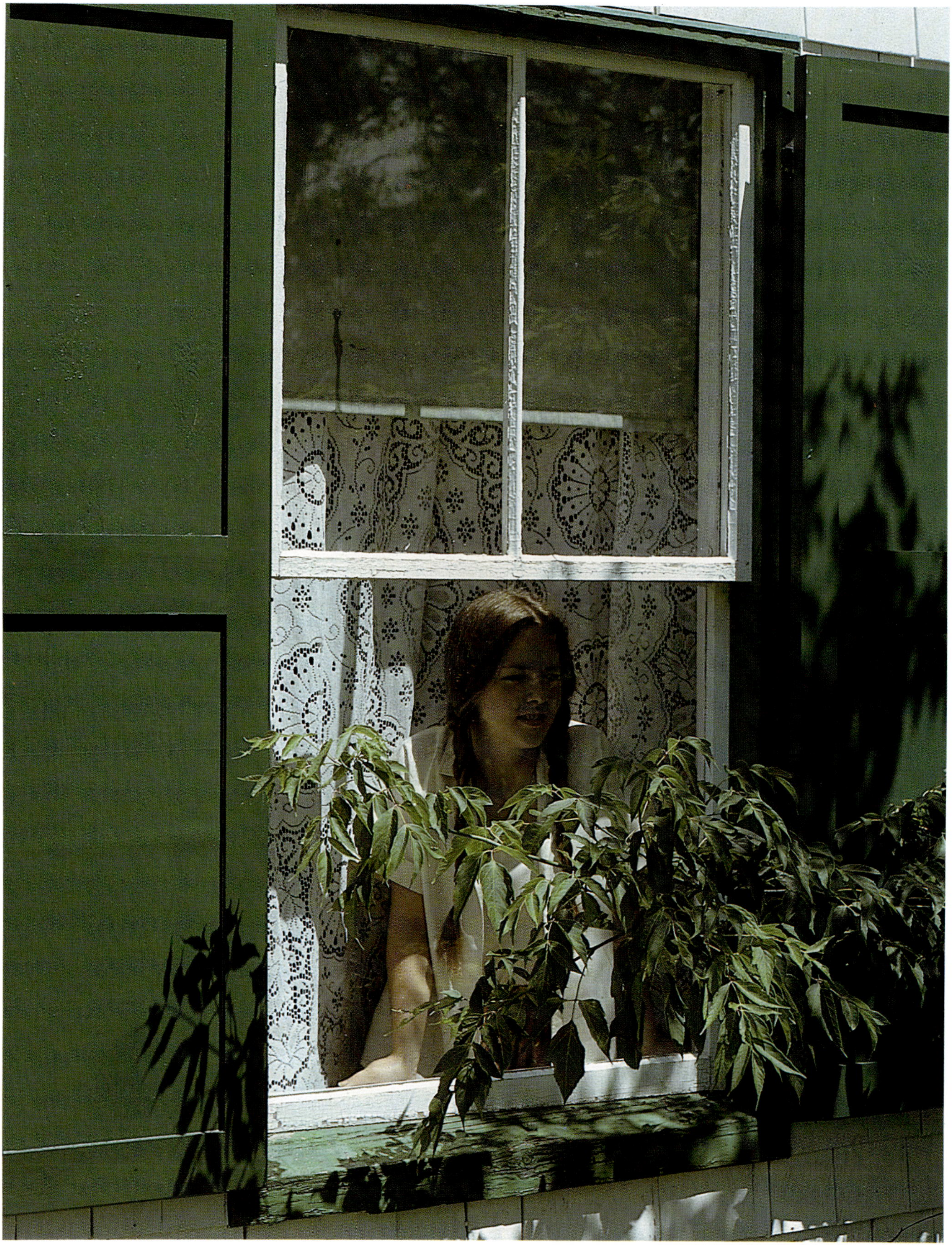

Green Gables, Cavendish

The magnetic appeal of Green Gables has to be traced to Lucy Maud Montgomery's own life in Cavendish. Shortly after Lucy Maud's birth in New London (Clifton), in 1874, her mother developed tuberculosis, soon to become so debilitating that her husband was obliged to move his wife and young child to the maternal homestead in Cavendish. After her mother's death in Cavendish and her father's decision to move to the West, it was decided that Lucy Maud would continue to make her home with her maternal grandparents, Alexander and Lucy Macneill.

Lucy Maud spent thirty-two of the thirty-seven years that she lived on the Island in her beloved Cavendish. There she published, in 1908, her first novel, *Anne of Green Gables*, which won international acclaim. In 1911, after writing three more novels, she married Rev. Ewen Macdonald, a Presbyterian minister. Although they lived in Ontario for the rest of their lives, the Island remained her spiritual home, illustrated by the fact that nineteen of her twenty novels are set, wholly or in part, on Prince Edward Island. She has often stated that the incidents and environment of her youth had a deep influence on her, and that, if it had not been for her years in Cavendish, *Anne of Green Gables* would not likely have been written.

The national park, established in 1936, appropriately included Cavendish as its western section. And in Cavendish the park comprised not only the narrow coastline strip but also the land surrounding the farmhouse known as Green Gables. Although Lucy Maud has always insisted that Green Gables was 'practically imaginary', there *are* real associations. Lover's Lane, the Haunted Wood, Dryad's Bubble, and the Birch Path, all of which figured prominently in Anne's life, were also favourite haunts of her creator. And the house now known as Green Gables, the former residence of her close friend Mrs. Ernest Webb, and earlier of her Macneill relatives, is quite representative of the farmhouses of the early 1900s when the first novel was written. Each year thousands experience the coming alive of *Anne of Green Gables* as they visit the frame farmhouse and walk through the fields and woods and along the sea-shore of Avonlea.

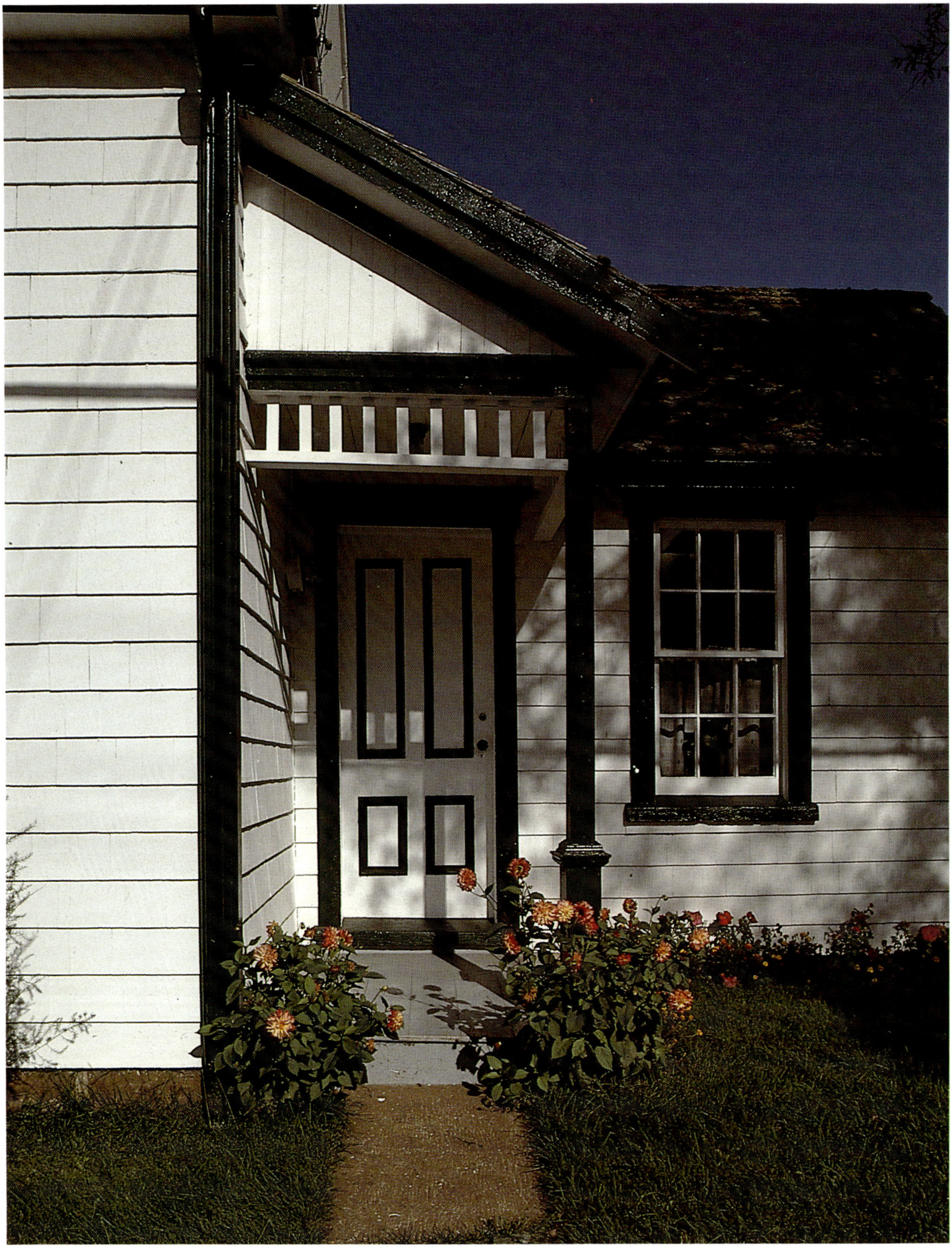

Lucy Maud Montgomery Birthplace, New London (left)

The birthplace of the world-famous author, Lucy Maud Montgomery, is in the small village of New London (Clifton), overlooking her beloved New London Harbour, the sand dunes, and the Gulf of St. Lawrence. Lucy Maud, daughter of Hugh John Montgomery and Clara Woolner Macneill, was born in this comfortable little house on 30 November 1874. Her father operated a general store, which he called 'Clifton House,' on the same property.

The birthplace, although passing through several hands over the years, remains relatively unchanged, still on the original location. The central attraction of this house, still tastefully furnished with Victorian pieces, is undoubtedly the bedroom in which the novelist was born. On display is a collection of twelve of her personal scrapbooks, depicting highlights of her life as a student and teacher, and copies of 396 of her stories as well as many poems published throughout her career. Her wedding dress and accessories are also on display. In 1964 the centenary of the Charlottetown Conference, K.C. Irving, who had owned the house for several years, generously deeded it to the province of Prince Edward Island. It is administered, operated, and maintained by the Lucy Maud Montgomery Foundation Board, appointed by the government of Prince Edward Island.

Hampton

Memories of the Old Home Place (right)

In modern Island society, quilts have been elevated from attractive household furnishings to colourful examples of folk art. To the talented and imaginative quilter, the quilt can become a drawing board upon which nostalgic memories may be graphically depicted. Ruth Paynter delineates the sentiments her handsome quilt evokes:

> *An old photo album often reveals the activities of family living over the years. Here is my 'quilted' album of memories.*

(overleaf) *Charlottetown*

85

BIBLIOGRAPHY

Books

The Agricultural Historians, *History and Development of the Dairy Industry in Prince Edward Island* (Charlottetown, 1978).

Alberton Historical Group, *Footprints on the Sands of Time: A History of Alberton* (Summerside, 1980).

Arsenault, A.E., *Memoirs of Aubin Edmond Arsenault* (Charlottetown, n.d.).

Arthur, E., and D. Witney, *The Barn: A Vanishing Landmark in North America.* (New York, 1972).

Baglole, H. and R. Irving, *The Chappell Diary: A Play of Early Prince Edward Island* (Charlottetown, 1977).

Birch Hill Women's Institute, *Climbing the Hill: A History of Birch Hill* (Summerside, 1982).

Blanchard, J.H., *The Acadians of Prince Edward Island, 1720-1964* (Québec, 1964).

Bolger, F.W.P. (ed.), *Canada's Smallest Province* (Toronto, 1973).

– *Prince Edward Island and Confederation, 1863-1873* (Charlottetown, 1964).

– *The Years Before 'Anne'* (Charlottetown, 1974).

Bolger, F.W.P., W. Barrett, and A. MacKay, *Spirit of Place: Lucy Maud Montgomery and Prince Edward Island* (Toronto, 1982).

Bolger, F.W.P., and E.R. Epperly (eds.), *My dear Mr. M: Letters to G.B. MacMillan from L.M. Montgomery* (Toronto, 1980).

Campbell, T. and R. Paynter, *From the Top of the Hill, The History of an Island Community Irishtown–Burlington* (Summerside, 1977).

Clark, A.H., *Three Centuries and the Island* (Toronto, 1959).

de Jong, N., and M. Moore, *Launched From Prince Edward Island* (Charlottetown, 1981).

Forrester, J.E., and A.D. Forrester, *Silver Fox Odyssey: The History of the Canadian Silver Fox Industry* (Charlottetown, n.d.).

Foster, Eidon, and Evelyn Foster, *French River and Park Corner History, 1773-1973* (1973).

Greenhill, B., and A. Giffard, *Westcountrymen in Prince Edward's Isle* (Toronto, 1967).

Hennessey, M. (ed.), *The Catholic Church in Prince Edward Island, 1720-1979* (Summerside, 1979).

Hickey, K., *Facts of St. Mary's Church, Indian River, Prince Edward Island From Beginning to Present* (1978).

Lothian, W.F., *A History of Canada's National Parks*, vol. 1 (Ottawa, 1976).

MacDougall, A., and V. MacEachern, *The Banks of the Elliott* (Charlottetown, 1973).

MacFadyen, *For the Sake of the Record* (Summerside, n.d.).

MacLennan, J.M., *From Shore to Shore* (Edinburgh, 1977).

MacMillan, J.C., *The Early History of the Catholic Church in Prince Edward Island, 1721-1835* (Quebec, 1905).

– *The History of the Catholic Church in Prince Edward Island From 1835 Till 1891* (Québec, 1913).

The Malpeque Historical Society, *Malpeque and Its People* (Summerside, 1982).

Martin, K., *Watershed Red: The Life of the Dunk River, Prince Edward Island* (Charlottetown, 1981).

Millman, T.R., *The Parish of New London* (Toronto, 1959).

Moase, M.L., *The History of New Annan, 1800-1971* (Summerside, 1971).

Pendergast, G., *A Good Time Was Had by All* (1980).

Pollard, J.B., *Historical Sketch of Prince Edward Island, Military and Civil* (Charlottetown, 1897).

Rankin, R.A., *Down at the Shore* (Charlottetown, 1980).

Rogers, I.L. *Charlottetown: The Life in Its Buildings* (Charlottetown, 1983).

Shaw, W., *Tell Me the Tales* (Charlottetown, 1975).

Sloane, E., *An Age of Barns* (New York, 1967).

Springfield Women's Institute, *Springfield, 1828-1953* (1954).

Street, D., *Horses: A Working Tradition* (Toronto, 1976).

Tuck, R.C., *Gothic Dreams: The Life and Times of a Canadian Architect: William Critchlow Harris 1854-1913* (Toronto, 1978).

Wallace, J., *The Eel Fishery Resource of Prince Edward Island* (1973).

White, P.C.T. (ed.), *Lord Selkirk's Diary, 1803-04* (Toronto, 1958).

Witney, D., *The Lighthouse* (Toronto, 1975).

Wood, K. (ed.), *An Historical Compilation of Victoria by the Sea, Our Legacy and Trust* (1973).

Pamphlets

Baird, D.M., *Prince Edward Island National Park, The Living Sands*, Miscellaneous Report 3 (Ottawa, Geological Survey of Canada, 1963).

Cayo, D. (ed.), *Green Gables* (Charlottetown, Parks and People Association Inc., 1982).

Institute of Man and Resources, *Atlantic Wind Test Site* (Charlottetown, 1983).

Parks Canada, *Ardgowan National Historic Park, Prince Edward Island* (Ottawa, 1982).

– *The Blue Heron* (Ottawa, 1978).

– *Dunelands. Prince Edward Island National Park, 1978* (Ottawa, 1978).

– *Province House* (Ottawa, 1980).

Potato Marketing Board, *Prince Edward Island Potato Handbook* (Charlottetown, 1980).

Prince Edward Island Heritage Foundation Government House Committee. *Government House* (Charlottetown, 1973).

Rogers, I.L. *Walks in Charlottetown* (Charlottetown, 1980).

Newspapers

The Examiner, 15 September 1902.

The Guardian, 30 August 1967.

The Guardian, 19 April 1969.

The Guardian, 23 May 1973.

The Guardian, 27 March 1976.

The Guardian, 10 August 1981.

The Guardian, 27 August 1981.

The Guardian, 3 April 1982.

The Guardian, 19 May 1982.

The Guardian, 30 July 1982.

The Guardian, 5 August 1982.

The Guardian, 8 October 1982.

The Guardian, 21 December 1982.

The Islander, Charlottetown, 5 April 1845.

The Journal Pioneer, Summerside, 26 May 1972.

The Patriot, 22 January 1876.

P.E.I. Agriculturist, 16 July 1925.

The Royal Gazette, Charlottetown, 16 December 1845.

West Prince Weekly, 9 August 1971.

Manuscripts

Cullen, M.K., 'A History of the Structure and Use of Province House, Prince Edward Island' (Ottawa, National Historic Parks and Sites Branch, 1977).

Gillis, G.M., 'Point Prim Lighthouse, Past and Present'. Research paper, University of Prince Edward Island, unpublished (Charlottetown, 1983).

Horne, F., 'Prince Edward Island National Park'. Manuscript Report No. 352 (Ottawa, 1978-9).

Public Documents

Journal of the House of Assembly of Prince Edward Island, Appendix B. (1845).

Journal of the House of Assembly of Prince Edward Island, Appendix M. (1847).

Parliamentary Reporter: Debates and Proceedings of the House of Assembly of Prince Edward Island for the Year 1873 (Charlottetown, 1873).

Articles

Bagnall, R., 'Roadhouses and Taverns', *Historic Highlights of Prince Edward Island* (Charlottetown, Historical Society of Prince Edward Island, 1955), 52-8, 109-15.

Bolger, F.W.P., 'Momentous Issues in the Island Story', *Canadian Antiques Collector*, viii (Prince Edward Island Centenary Issue No. 1, 1973), 9-12.

Brehaut, M., 'The Selkirk Pioneers', *Historical Highlights of Prince Edward Island* (Charlottetown, Historical Society of Prince Edward Island, 1955), 32-7.

Bumstead, J.M., 'Lord Selkirk of Prince Edward Island', *The Island Magazine*, Fall-Winter (1978), 3-8.

Bush, E.F., 'The Canadian Lighthouse', *Canadian Historic Sites: Occasional Papers in Archaeology and History*, no. 9 (Ottawa, Parks Canada, Indian and Northern Affairs, 1974), 55.

Cairns, W., 'On the North Shore', *Nature Canada*, April-June (Ottawa, 1980), 47-53.

Callbeck, L.C., 'A Faithful Friend—the Potato', *The Atlantic Advocate*, April (1967), 55-6.

Campbell, J.C., 'School Consolidation: Seventy Years of Evolution', *The Abegweit Review*, Fall (1975), 1-10.

Cousins, J., 'Those Gallant Ships', *Canadian Antiques Collector*, viii (1973), 58-60.

Croteau, J.T., 'The Farmers' Bank of Rustico: An Episode in Acadian History', *The Island Magazine*, Spring-Summer (1978), 3-8.

Fisher, L.R., 'The Shipping Industry of Nineteenth Century Prince Edward Island: A Brief History', *The Island Magazine*, Spring-Summer (1978), 15-21.

Goldsmith, O., 'The Deserted Village', *The Norton Anthology of English Literature* (New York, 1974), 2405-14.

Griffin, D. 'Prince Edward Island Sand Dunes', *The P.E.I. Environer*, Summer (1974), 4.

Griffin, D. and I. MacQuarrie, 'The Living Sands, Coastal Dunes of P.E.I.', *Nature Canada*, April-June (1980) 42-6.

Harkness, D. 'The Straightaway and Home', *Axiom*, July-August (1977), 14-6.

Hennessey, C.G., 'Heritage Foundation', *Canadian Antiques Collector*, viii (Prince Edward Island Centenary Issue No. 1, 1973), 66-7.

Hornby, J. 'The Great Fiddling Contests of 1926', *The Island Magazine*, Fall-Winter (1979), 25-30.

Horne, H.S., 'The Fair That Everyone Loved', *The Atlantic Advocate*, October (1977), 66-9.

MacRae, A., 'North Cape, Light Long Awaited', *West Prince Graphic*, 18 October 1981.

Rankin, R.A., 'Robert T. Oulton and the Golden Pelt', *The Island Magazine*, Fall-Winter (1977), 17-22.

Robertson, I.R., 'George Coles', *Dictionary of Canadian Biography, Vol. X: 1871-1880*, 182-8.

- 'William Henry Pope', *Dictionary of Canadian Biography, Vol. X: 1872-1880*, 593-7.

Ross, L.M., 'Houses of Worship', *Canadian Antiques Collector*, viii (Prince Edward Island Centenary Issue No. 1, 1973), 36-9.

Stewart, C.B., 'Roadblock 1810', *The Island Magazine*, Spring-Summer (1983), 14-8.

Thomas, V., 'Gourmet Holidays', *Gourmet Magazine*, May (1977), 99.

Tuck, R.C., 'Victoria: Seaport on a Farm', *The Island Magazine*, Winter (1979), 38-44.

- 'William Harris and His Island Churches', *The Island Magazine*, August (1899), 220-4.

Watson, L.W., 'Among Our Orchids', *The Prince Edward Island Magazine*, August (1899), 220-24.

Weale, D., 'The Minister': The Reverend Donald McDonald', *The Island Magazine*, Fall-Winter (1977), 1-6.